VIDEOGAMES

James Newman

Routledge
Taylor & Francis Group

LONDON AND NEW YORK

The only legitimate use for a computer is to play games

Eugene Jarvis,
videogame designer

First published 2004
by Routledge
11 New Fetter Lane, London EC4P 4EE

Simultaneously published in the USA and Canada
by Routledge
29 West 35th Street, New York, NY 10001

Routledge is an imprint of the Taylor & Francis Group

© 2004 James Newman

Typeset in Perpetua and Univers by
Florence Production Ltd, Stoodleigh, Devon
Printed and bound in Great Britain by
MPG Books Ltd, Bodmin, Cornwall

British Library Cataloguing in Publication Data
A catalogue record for this book is available from
the British Library

Library of Congress Cataloging in Publication Data
A Catalog record for this book has been requested

ISBN 0–415–28191–1 (hbk)
ISBN 0–415–28192–X (pbk)

CONTENTS

TABLES

SERIES EDITOR'S PREFACE

There can be no doubt that communications pervade contemporary social life. The audio-visual media, print and other communication technologies play major parts in modern human existence, mediating diverse interactions between people. Moreover, they are numerous, heterogeneous and multi-faceted.

Equally, there can be no doubt that communications are dynamic and ever-changing, constantly reacting to economic and popular forces. Communicative genres and modes that we take for granted because they are seemingly omnipresent – news, advertising, film, radio, television, fashion, the book – have undergone alarming sea changes in recent years. They have also been supplemented and reinvigorated by new media, new textualities, new relations of production and new audiences.

The *study* of communications, then, cannot afford to stand still. Although communication study as a discipline is relatively recent in its origin, it has continued to develop in recognizable ways, embracing new perspectives, transforming old ones and responding to – and sometimes influencing – changes in the media landscape.

This series of books is designed to present developments in contemporary media. It focuses on the analysis of textualities, offering an up-to-date assessment of current communications practice. The emphasis of the books is on the *kind* of communications which constitute the modern media and the theoretical tools which are needed to understand them. Such tools may include semiotics (including social semiotics and semiology), discourse theory, poststructuralism, postcolonialism, queer theory, gender analysis, political economy, liberal pluralism, positivism (including quantitative approaches), qualitative methodologies (including the 'new ethnography'), reception theory and ideological analysis. The breadth of current communications media, then, is reflected in the array of methodological resources needed to investigate them.

Yet, the task of analysis is not carried out as an hermetic experiment. Each volume in the series places its topic within a contextual matrix of production and consumption. Each allows readers to garner an understanding of what that communication is like without tempting them to forget who produced it, for what purpose, and with what result. The books seek to present research on the mechanisms of textuality but also attempt to reveal the precise situation in which such mechanisms exist. Readers coming to these books will therefore gain a valuable insight into the present standing of specific communications media. Just as importantly, though, they will become acquainted with analytic methods which address, explore and interrogate the very bases of that standing.

WHY STUDY VIDEOGAMES?

TAKING GAMES SERIOUSLY

1962 was an auspicious year. Lt. Col. John Glenn, Jr became the first American to orbit the Earth, Telstar became the first communications satellite intended for regular service relaying television signals between North America and Europe, Marilyn Monroe died from a suspected overdose of sleeping pills, President John F. Kennedy found himself at the centre of the Cuban Missile Crisis, and videogames were born. When Steve Russell created *Spacewar* in the computer labs of the Massachusetts Institute of Technology, he started a chain of events that would change not only computing, but also entertainment and popular culture – not immediately, and certainly not knowingly, but decisively and permanently.

Two things might already surprise even videogame aficionados. The first is that videogames are over 40 years old. In fact, their birth date is debated and, while *Spacewar* is the most oft-cited 'modern' computer game, there remains no absolute consensus, with some commentators citing the 1958 *Tennis for Two* as the true original. This uncertainty, as we shall learn in Chapter 2, is revealing, as the argument as to what constitutes 'a videogame' consumes much of the effort of scholars in the field. The second surprise is the suggestion that something as apparently trivial as videogames should be taken so seriously. While there can be few who would suggest that they are as significant as manned spaceflight or possible world war, a growing number of scholars and cultural critics are coming to recognize the social, cultural and economic importance of this form of entertainment. Henry Jenkins (2000) proclaims videogames 'an art form for the digital age'. Taking inspiration from Gilbert Seldes who, in the 1920s, championed Hollywood movies, jazz, comic strips and Broadway musicals, Jenkins suggests that videogames must be considered to be one of the most important artforms of the twentieth century. Ralph Koster similarly implores us to consider computer games as art:

> notice how much scorn gets heaped on games that are perceived as mere clones or knockoffs. The public already discusses and treats games as an art form, and uses the same standards of judgment for them as they do for films or novels or any other artistic medium. They just aren't used to considering them to be art.
>
> (Koster 1999)

Shuker (1995) is even more enthusiastic, 'Video games are now a major cultural form, and may well soon replace cinema, cable and broadcast television as the dominant popular medium'. We might still be some way from seeing Shuker's prediction come true, but even if, as Jesper Juul (2000) has more pragmatically suggested, we have not seen the first videogame Shakespeare or Bach, the speed with which videogames have developed aesthetically, formally and functionally, is remarkable. The level of audio-visual and interactional sophistication of today's Play-Station 2, Xbox, GameBoy Advance and GameCube games seems light years away from 1970s' offerings like *Space Invaders*, *Pac-Man* and *Asteroids*, though, as we will learn throughout this book, there remain considerable areas of constancy and clearly identifiable lineages.

However, notwithstanding the developments in the aesthetic capabilities of videogame systems, or the ever more sophisticated interface technologies employed in both the home and arcade via which players engage with the gameworld, is it really appropriate to suggest that videogames are *that* important? Surely, compared with television or film, videogames are mere trivia, just a passing fad. Worse still, the games might appear impenetrably complex and monotonous. As Jessen (1998) has pointed out, it seems almost inconceivable that players could have spent so long absorbed by games as repetitive and simplistic as *Pac-Man*. To the untrained eye, videogames are as incomprehensible as abstract art or experimental music. Jessen encourages us to consider Stanley Fish's concept of 'interpretive communities' in appreciating the pleasures of videogames. Not only can the game not be understood without playing it, but also the game is only part of the experience. Meaning is not embedded within the game, but rather is revealed through use.

Consequently, the context in which the game is used or played affects and shapes its value:

> You have to play the game before it will reveal its nature, and this is something that far from happens to everyone. Some fall for it, others find it monotonous, boring and pointless, but whatever attitude one has to the game, the interest rarely lasts long. If one plays it individually, Pac-Man may be exciting at first, but rather boring in the long run. The game only takes on a content in a social context.
>
> (Jessen 1998: 38)

The lack of sensitivity to the experiences of play and the use of videogames by players is perhaps one of the most serious deficiencies in extant scholarly studies and is a subject to which this book turns its attentions frequently (see Chapters 4 and 9 especially).

WHY STUDY VIDEOGAMES?

While scholars identify a range of social, cultural, economic, political and technological factors that suggest the need for a (re)consideration of videogames by students of media, culture and technology, here, it is useful to briefly examine just three reasons why videogames demand to be treated seriously: the size of the videogames industry; the popularity of videogames; videogames as an example of human–computer interaction.

The global videogames industry is enormous. In 1999, UK players spent almost £1 billion on computer and videogames and for the first time, spending on videogames outstripped video sales. According to ELSPA (European Leisure Software Publishers Association) sales of videogames in 2001 totalled £1.6 billion. This 36 per cent increase in the demand for software and hardware meant that 'People in the UK now spend more money on computer games than on renting videos or going to the cinema' ('Record year for computer games', BBC News Online 2002). Impressive though these figures are, it is worth noting that the UK is only the third largest market behind the US and Japan respectively. 'US games sales were a record $9.4 billion last year – topping the previous 1999 record by about $3 billion, and exceeding the Hollywood box office by $1 billion' ('US sales go big', *Digitiser* 2002: 481). Videogames are big business and getting bigger. At the 2002 Electronic Entertainment Expo (E3), market leader Sony announced global shipments of 30 million PlayStation 2 consoles (adding to installed base of over 89 million PSone consoles). In PAL (Phase Alternate Line) territories (predominantly Western Europe) Sony announced that they were selling 70,000 PlayStation 2s a week, with

30,000 PSones. By March 2003, projections suggest that some 13.5 million PlayStation 2s will be installed in Europe alone.

Considering the sales figures above, it might seem unnecessary to claim that videogames are popular. However, the degree of popularity and their relation to other media is surprising. Sherry *et al.* (2001: 1) note, 'In 2000, 35 per cent of Americans identified video games as the most fun entertainment activity (second was television at 18 per cent)'. In addition to consuming enormous amounts of money, videogame play also consumes enormous amounts of time. As we will learn in Chapter 4, not only do players play frequently – usually many times a week and often every day – but individual play sessions are typically long. Uninterrupted play sessions may last many hours. It is notable also, and indicative of the high levels of engagement and involvement with videogames, that play sessions are almost always longer than players intend or even realize. Because videogames can be absorbing, and because designers deliberately build in 'hooks' to encourage players to replay and revisit their games, and because this is demanded by players who privilege longevity and replayability in videogames (see Juul 1999), some detractors have gone as far as to label videogames 'addictive', even comparing them to drugs (see Grossman 2001, for example).

If videogames are bad for players as Grossman and others suggest, then the large amounts of time spent engaging with them must be a cause for concern. However, it is worth considering that videogames may have some positive benefits. As we will learn in Chapter 6, videogames are quite possibly the most sophisticated and certainly most pervasive example of high-level human–computer interaction presently available and, as such, provide a useful environment to learn about, and become proficient with, technology. Consider, for example, the amount of time it takes to learn how to use an average office productivity tool such as a word processor e-mail package – not to become an expert, just to get a feel for the program and acquire a basic competence. Then compare this with the amount of time it takes to get up and running with the average videogame. Watch a player encountering a brand new game and, after an initial, quite inevitable, but surprisingly short, period of acclimatization, you will doubtless find a player performing complex series of immensely precise and intricate interactions with little apparent effort and certainly minimal, if any, contemplation of the joypad or other physical input devices. Perhaps the player is not conquering the game straight away. Perhaps they are not even making as much progress as they think they should be. Perhaps the controller has been thrown across the room in disgust a few times and, as Jones (1997) notes, failure is an essential feature of both games and learning. However, what is almost certain is that (once it has been picked up again), that controller will look like an extension of the player's hand after just a few minutes of play. Consider also that this player may well be a pre-schooler and the significance of videogaming as a mechanism through which a comfort and familiarity with computers can be engendered becomes clear. Not only can players learn from videogames, but designers and developers of 'serious' computing applications could surely benefit from an examination of the interfaces of games like *Super Mario 64* (see Livingstone 2002; Kafai 2001; Kasvi 2000; Amory *et al.* 1998; Jones 1997; Seay 1997; Leyland 1996).

WHY HAVE ACADEMICS IGNORED COMPUTER GAMES?

Given the clear impact of videogames and videogame play in everyday life, it is surprising to note that academics, particularly scholars of media and cultural studies, have largely neglected them. Those scholarly studies that do exist chiefly emanate from the research laboratories of psychology departments and are typically concerned with the possible effects of games on young players. Moreover, a good deal of these studies date from the mid- to late-1980s. As Jonas Smith (2001a) has noted, videogames are essentially a forgotten medium. There are a number of possible reasons. Here, we will concentrate on two misconceptions. First, videogames are seen as being a children's medium. This means that they are easily and readily denigrated as trivial – something that will be 'grown out of' – and demanding no investigation. Second, videogames have been considered mere trifles – low art – carrying none of the weight, gravitas or credibility of more traditional media. As a consequence, they have been unfavourably and unfairly, compared to respectable media like film and even parts of the videogames industry seem embarrassed about their business (see Sheff 1993: 1, for example).

Simply dismissing the *Super Mario Bros* series as childish because of its representational style (bold, primary colours) or even the apparent nature of their content (jumping on enemies' heads) reveals a superficiality in the investigation of the games. Even a cursory survey of fan comments and reviews, reveals that games such as *Super Mario Kart* are revered by players young and old because of their impeccable balance and feel (see player reviews on fansites such as www.gamefaqs.com, for example). In fact, many players rate *Super Mario Kart* among their favourite games *despite* the representational style (Newman 2001). In this way, a concentration solely on the appearances of the game may reveal a lack of appreciation of the experience of play and the preferences and motivations of videogame players (see Rollings and Morris 2000). The relegation of the 'childish' is immensely naive in its own right, though it is perhaps unsurprising given the pervasive Western tradition of denigrating the anthropomorphism that features so heavily in videogames such as *Super Mario Bros* and *Sonic the Hedgehog*, by associating it with immature, childish thinking (see Piaget 1929, for example). However, it is particularly ironic to note that the relegation of videogames to childish plaything reveals a fundamental misunderstanding of the contemporary marketplace and audience demographic. The effect of aggressive marketing campaigns employed by Sony to promote PlayStation and PlayStation 2 has been a shift in videogame market demographics with the average age of players continuing to rise year on year. From the outset, the PlayStation family was marketed at late teens/early twenties and the success of these strategies has contributed to a shifting demographic that must force a reconsideration of the videogame as merely a child's toy. In their 2001 industry overview report, the Interactive Digital Software Association (IDSA) claim that the current average age of US videogame players is 28 ('Top ten industry facts' 2001; see also Chapter 4).

The more general denigration of 'mere entertainment' is partly responsible for the lack of seriousness in the treatment of videogames, 'One of the commonest points I hear about why video games are not an art form is that they are just for fun. They are just entertainment' (Koster 1999). The deprecation of 'popular' and particularly 'youth' culture has been well documented, for example, in the work of the Centre for Contemporary Cultural Studies. Where it has been researched, youth culture and its associated genres and communicative forms have frequently been presented as potentially dangerous, with studies typically focusing on deviance and resistance (cf. Skelton and Valentine 1998; McRobbie and Thornton 1995; Skirrow 1990; Cohen 1972, 1956). As we will learn in Chapter 4, many of the extant research studies investigating computer and videogames have centred on the potentially damaging and antisocial effects of play (see Griffiths 1999; Dill and Dill 1998 for reviews). This lack of scholarly engagement should surprise us little and videogames can by no means be considered unique in this regard. As Roger Sabin (1998) has noted, comics have been similarly overlooked as a form worthy of serious study. More surprising, perhaps even more significant, is that it seems possible to identify a similar unease, almost embarrassment, about videogames within certain quarters of the industry itself. Most notably, it is not even universally referred to as the 'videogame' industry. Various euphemisms have passed into common parlance, all seemingly motivated by a desire to avoid the use of the word 'game' and perhaps even 'computer', thereby adding a veneer of respectability, distancing the products and experiences from the childish pursuits of game, play and toys, and downplaying the technology connection with its unwanted resonances of nerds in bedrooms hunched over ZX Spectrums and Commodore 64s and the amateurism of hobbyist production (see Davies 2001 on changing development methodologies). Thus, we find certain companies preferring to consider themselves contributing to a world of 'interactive entertainment'. *Edge* magazine is one of the UK's (and the global industry's) most respected publications, yet even here 'game' seems a dirty word; the magazine's strapline reading 'The Future of Interactive Entertainment'. It follows that the products of an interactive entertainment industry are not games. Rather, they are 'interactive fiction' or 'interactive narratives' (Juul 1999).

2

WHAT IS A VIDEOGAME? RULES, PUZZLES AND SIMULATIONS

Defining the object of study

SUPER MARIO, TAMAGOTCHI, FURBY AND AIBO

We have noted that there is a surprising confusion among consumers, producers and scholars of videogames as to just which experiences or products constitute 'a videogame'. Part of the problem doubtless arises from the enormous variety of game types that come under the broad umbrellas of 'videogames', 'computer games' or 'interactive entertainment'. Prima facie, it is difficult to spot immediate similarities between a game with no graphics, no sound and text-only input played on a home computer and a flight combat game like Sega's *R360* in which the player is strapped into a mock-up fighter plane cockpit, clutches a force feedback joystick, and is literally thrown around as the entire game cabinet turns side-to-side and upside-down in correspondence with the action in the gameworld. The dissimilarities between a 'bemani' ('beatmania') game in which the player is required to physically input dance steps on a pressure-sensitive playmat/dancefloor and a word puzzle game played on a mobile phone seem far more obvious than the similarities. As Aarseth (2001a) has asked, is Furby a computer game? It is certainly computer-controlled and we play with it. What about Tamagotchi? Or Sony's AIBO . . . ?

The lack of in-depth studies has given rise to a situation in the study of videogames in which the definition and demarcation of the object of study is a matter of debate. This highlights an important point at the centre of game and digital culture studies. It is perfectly possible to conduct a thorough discussion of computer and videogames with no consensus as to precisely what forms, experiences, or technologies are under examination. Should we see videogames as continuations of other media such as film or television? Are they continuations of other, non-computer, games? Are they hybrids of both? Should we define them with reference to their uniqueness and dissimilarity from other entertainments, media or games, or as a consequence of their similarity? As we will learn throughout this book, the answers to these and other questions will be at least partly shaped by the experience and biography of videogame theorists and critics themselves. In 2001, at the first academic videogames conference in the UK, Henry Jenkins conceived the problem in terms of a number of blind people attempting to describe an elephant. For one, the elephant is all about the trunk, while for the others, it is the tail, or the ears . . . For scholars of film theory, it is perhaps natural to view videogames as forms of 'interactive narrative', for example, while for scholars of play and games, they will be understood very differently. These competing approaches have already created a schism in videogame studies between narratologists and ludologists (see Chapter 6). However, it is clear that at this comparatively early stage in the evolution of videogame studies, no single group of theorists can claim to be able to accurately describe the elephant. It is perhaps for this reason that Aarseth (2001b) has proclaimed 2001 'year one of the discipline'.

Given that the industry cannot even agree on a name for its products and scholars see different characteristics as important perhaps depending on their academic background or the games they study, just how does one start to home in on what a videogame is? Geoff Howland (1998a) breaks down the videogame into five distinct, yet interconnected, elements (see Table 2.1).

While such distinctions are useful, they are too broad to capture the array of videogame experiences. The inclusion of 'graphics' implies the centrality of a screen thereby prohibiting the consideration of AIBO, Furby and also text-only adventures, but potentially encompassing Tamagotchi or Digimon (see also Livingstone 2002 on videogames as 'screen based entertainment media'). Furthermore, if the categories are problematic in delimiting videogames from other media forms, they are of even less use in distinguishing videogames from one another. We will temporarily leave aside this first issue of distinguishing videogames from other media, or other games, and turn our attentions to attempts to delineate different types of videogame.

Table 2.1 The elements of a videogame

Graphics	Any images that are displayed and any effects performed on them. This includes 3D objects, 2D tiles, 2D full-screen shots, Full Motion Video (FMV), statistics, informational overlays and anything else the player will see.
Sound	Any music or sound effects that are played during the game. This includes starting music, CD music, MIDI, MOD tracks, Foley effects, environmental sound.
Interface	The interface is anything that the player has to use or have direct contact with in order to play the game . . . it goes beyond simply the mouse/keyboard/joystick [and] includes graphics that the player must click on, menu systems that the player must navigate through and game control systems such as how to steer or control pieces in the game.
Gameplay	Gameplay is a fuzzy term. It encompasses how much fun a game is, how immersive it is and the length of playability.
Story	The game's story includes any background before the game starts, all information the player gains during the story or when they win and any information they learn about characters in the game.

Source: Adapted from Howland 1998a.

CLASSIFYING VIDEOGAMES

The seemingly bewildering variety of game types renders it almost inevitable that computer and videogame theorists, journalists and marketers have attempted to find ways of classifying and making more manageable the object of their attentions. By far the most frequently used tool has been genre. The generic classification of computer and video-games is so widely employed that it is often easy to overlook it altogether or merely consider it natural. References to 'shoot-'em-ups', 'driving games', 'platform games' abound in videogame review magazines while academic research projects frequently orient their analyses around similar classifications (see Griffiths 1999, 1997a, for example). The terminology of 'shoot-'em-up' and 'beat-'em-up' appears to have originated in the mid- to late-1980s in the pages of early games magazines such as Newsfield's *Crash* and *Zzap!64* and has now passed into industry parlance. In their review of the best games for 2002's gaming platforms, Berens and Howard (2001: 25–26) demonstrate the continued relevance of industry-derived genres. 'they are useful pointers and reflect the industry's current view of how they operate (it's also how you may well find them organized in the bigger retailers)'. Conflating some similar categories, Berens and Howard present seven game types:

1   Action and Adventure
2   Driving and Racing
3   First-Person Shooter
4   Platform and Puzzle
5   Roleplaying
6   Strategy and Simulation
7   Sports and Beat-'em-ups.

However, the deployment of these genres in scholarly study may be problematic for a number of reasons. First, the categories are extremely nebulous and do not represent the fixities that the commentators that utilize them imply. Second, their use perhaps implies not only an overly text-centred approach to understanding videogame play but one in which the text is considered a hermetic, closed system (cf. Bakhtin 1981). Studies that seek to evaluate the effect or consequence of one game type in comparison with another necessarily divert attention from the location of play and players within specific socio-cultural, historical, and even interactional or 'ludic' contexts. In doing so, they perhaps betray a desire to decry the videogame as formulaic and the player as unsophisticated dupe (see the discussion of videogame effects research and the methodological challenges offered by Emes (1997) among others in Chapter 4, for example). Following Fish's (1980) work on interpretive communities, Bennett's 'reading formations' (1990, 1987) and the subsequent shift away from unveiling fixed, 'in-the-text' meanings that has accompanied the wealth of investigations into literary, film and television audiences (see Ang 1996, 1985; Hermes 1995; Morley 1986, 1980; Radway 1984, for example), it is possible to understand these genres in terms of the ways they cue or represent expectations harboured by the audience (see Cobley 2001b). Furthermore, as we shall learn in Chapter 9, the work of Jenkins (1992) and Lewis (1992) among others, points to the participatory nature of media consumption and the mutability of the text.

It is worth noting also that a recent trend in the marketing and criticism of games has seen the emergence of something akin to auteurism. Within the pages of review magazines, on fan websites and discussion boards and in scholarly research articles, the names of certain videogames designers and producers dominate. Perhaps most well-known and lauded is Shigeru Miyamoto, creator of some of Nintendo's most famous series including *Super Mario Bros* and *The Legend of Zelda*, for example. Similarly, Hideo Kojima (*Metal Gear Solid*), Yuji Naka (*Sonic the Hedgehog*), Sid Meier (*Civilization*) are discussed in terms of their unique inventiveness. Each of the two titles in the *Metal Gear Solid* series are publicized (both in advertising and marketing materials and within the game) as 'a Hideo Kojima game'. However, it is by no means clear whether the influence of these celebrity designers is felt aesthetically, formally or through the resultant 'feel' of the game, or whether the allusion to film is simply stylistic and indicative of a further attempt to gain credibility by association with 'respectable' media. Nevertheless, the esteem in which these apparently gifted designers are held provides considerable evidence that critics and players privilege both the innovative and challenging alongside the familiar and conventional. However, as we will learn in Chapter 3, the contemporary videogame industry is one in which commercial considerations ensure a raft of derivative, barely distinguishable sequels and it is perhaps for this reason that the 'celebrity designer' has been pushed to the fore. Whether there is something truly identifiable about a Miyamoto or Kojima creation is perhaps of secondary importance to the more general, and perhaps illusory, sense in which these designers are afforded a creative freedom to operate without constraint. Discussing the success of *Pokémon*, Miyamoto locates himself and the game's creator as artists, with commercial success an almost inevitable side-effect of the perfection of the artwork:

> the biggest reason it has become that popular is Mr Tajiri, the main developer and creator of *Pokémon*, didn't start this project with a business sense . . . Somehow, what he wanted to create for himself was appreciated by others in this country and is shared by people in other countries . . . And that's the point: not to make something sell, something very popular, but to love something, and make something that we creators can love.
>
> (cited in Rouse 2001: 1)

LUDIC CONTEXT: COIN-OP VS HOME SYSTEMS

An underused means of differentiating types of videogames and, more importantly, types of experience, structure and engagement, centres on the location of play. The overwhelming majority of studies of videogames concentrate, often exclusively, on home consoles or PCs with little, if any, consideration of arcade systems (though note Saxe 1994 on the social play in arcades). The distinction is important for a number of reasons. Coin-op systems are required to fulfil very different functions. At least one function of the machine is to earn money, therefore throughput is an important consideration. Among other things, including the physical design of the cabinet to ensure ease and speed of access, this pressure affects the nature of the experience to be delivered. Where a home console title may boast many tens of hours of gameplay (e.g. *Final Fantasy* series), requiring many consecutive play sessions, an encounter with an arcade title typically lasts just a few minutes. Even where an arcade game offers 'continues' (allowing players to resume their game from the point they exited rather than from the start), to continue demands depositing another coin, and each individual play session remains short. Consequently, the arcade experience is frequently characterized as one of sensory bombardment with intense, high volume and velocity play.

That videogames designed for the home may include an 'arcade' mode speaks of the variety of ways in which videogames are played and used. *Gran Turismo 2*, for example, distinguishes itself from other car racing games by the complexity and thoroughness of its options and the degree of personalization and customization it offers the player. Cars can be tuned, traded and upgraded with a multitude of extras, yet the 'Quick Arcade' mode, presents none of this intricacy and simply allows players to race around tracks in souped-up cars that would usually have to be painstakingly earned through a series of championships and the accumulation of credits. Clearly, the various modes of play serve different purposes. Engagement with the intricate 'Gran Turismo' requires a significant investment of time whereas 'Quick Arcade' mode not only satisfies the player with less time to dedicate to the game but also facilitates social play as players can alternate 3-lap sessions or challenge each other on the same screen via the 'Quick Arcade' mode's two-player simultaneous, split-screen option. In 'Quick Arcade' mode the pleasure of *Gran Turismo 2* is found in the honing of technique and the mastery of tracks and vehicles.

Another important area distinguishing arcade systems from home consoles and PCs is in the use of custom cabinets and interface technologies. *Virtua Cop*, for example, is particularly notable in employing a 'light gun' rather than standard joystick/button arrangement. The mediating technology in this case is perhaps a more significant differentiating characteristic than the content of the game itself. Even though light guns, steering wheels and even dance mats are available (though typically poorly supported) for use in the home, interfaces such as that found in *Alpine Skier*, *Final Furlong*, etc. clearly add a level of physicality to the gaming experience that is not recreated in conversions of the titles to console formats. Consider, too Sega's *Daytona USA* in its coin-op and home console incarnations. Both are driving games – in fact, both games are essentially identical – yet in the home you drive with your joypad, seated in your armchair, while in the arcade you sit in a mock-up car chassis, grasping a force-feedback steering wheel with pedals under your feet. This is certainly not intended to imply that home console games/ conversions cannot create exciting and captivating experiences that engage the player physically. However, it should be clear to even the least experienced scholar of computer and videogames that, if we compare pressing buttons on a standard, generic joypad with riding a virtual horse, the processes of engagement and interaction are likely to be different.

Quite apart from these differences, the actual location plays a role. While the home console game is likely to be played if not alone then among a group of family or friends known to the player, coin-op play is likely to take on at least an element of public performance as observers crowd around the machine. This crowding does not merely signal vicarious pleasure, but is one of the ways in which techniques and tactics are learned (Livingstone 2002). Moreover, as coin-op machines require money per play it makes sense to learn from somebody else's mistakes, at their expense. While 'home' and 'coin-op' are still too broad, the differentiation according to type of experience is a useful and important departure from content-derived genre classifications.

WHAT A VIDEOGAME ISN'T

As the variety of experiences and technologies makes it hard to define a videogame in positive terms, it is useful to challenge a number of preconceptions. For Rollings and Morris:

> a game is *not*:
> - A bunch of cool features
> - A lot of fancy graphics
> - A series of challenging puzzles
> - An intriguing setting and story
>
> (Rollings and Morris 2000: 19–20, original formatting)

It is important to note that videogames do not preclude these characteristics, rather that these qualities do not, in themselves, make a videogame or help us describe the uniqueness of the form. Graphics are clearly important, and we will learn in later chapters that the audio-visual composition of the gameworld has an enormous impact on players, non-players and purchasers of games alike. However, even the most aesthetically advanced gameworlds can fail as videogames. *Dragon's Lair* is a good example, though *Myst* has been claimed by many to fall into the same category in offering lavish visuals with minimal opportunities for interaction (Juul 1999). Moreover, *Tetris*, *Pac-Man*, *Vib Ribbon* all offer engaging, absorbing gaming experiences with minimal audio-visual flair while the GameBoy has become the best-selling videogames console despite its comparatively primitive audio-visual capabilities. Combining finely scripted narratives with cunningly ingenious puzzles does not necessarily make for the most rewarding gaming experiences (see Rollings and Morris' 2000: 22–23 discussion of *Baldur's Gate* and *Grim Fandango*, for example). Throughout this book, we will see discussion turn to each of these issues, particularly regarding the importance (or otherwise) of narrative, characters and aesthetics. For now, however, it is sufficient to conclude that wrapping fancy graphics around a narrative with intermittent impasses requiring input from the player does not make a videogame irrespective of the number of features that might never before have been seen.

WHY DO PLAYERS PLAY?

So, what do players want from videogames? Rouse (2001: 2–19) identifies a range of player motivations and expectations. Among them, three are particularly notable:

- challenge
- immersion
- players expect to do, not to watch.

Livingstone presents similar findings:

> In interviews with children regarding their experience of screen entertainment culture, what is most notable when children talk about *computer games*, the words that appear over and over are 'control', 'challenge', 'freedom'.
>
> (2002: 231)

All of these point to the importance of *player activity*. A videogame must provide novel or exciting situations to experience, stimulating puzzles to engage with, and interesting environments to explore. Moreover, it must offer the player not merely suitable or appropriate capabilities, but capabilities that can be earned, honed and perfected. Sherry *et al.* (2001) note also the importance of *challenge* in their study of videogame uses and gratifications. It is essential to note that players want to work for their rewards. Gratification is not simply or effortlessly meted out. As Rouse notes, players expect to fail. As we noted with *Gran Turismo 2* (see p. 14), at least part of the pleasure of videogame play is derived from the refinement of performance through replay and practice. Consequently, it is essential that obstacles, irrespective of the form they take, must be 'real' in that they must require non-trivial effort to conquer them. According to Rouse (2001), it is the primacy afforded to doing and performing that renders 'non-interactive cut-scenes' so unappealing to gamers. Cut-scenes are sequences in which the player is offered no direct control through the game's interface. Commonly referred to as 'intermissions' or more problematically as 'movie sequences', they typically introduce or frame 'gameplay' sequences or episodes in which direct control or 'interactivity' is offered (see Chapters 5 and 6 for more on the (in)significance of cut-scenes). Rouse makes the interesting observation that, while players demand participation and seem to tire quickly of non-participative elements, they want all of this presented in manner that does not feel contrived – in fact, in a manner that does not feel like a game at all:

> Once a player is into a game, she is in a level, she has a good understanding of the game's controls, she is excited, and she is role-playing a fantasy; she does not want to be snapped out of her experience . . . the player does not want to think about the game's GUI [graphical user interface]. If the GUI is not designed to be transparent and to fit in with the rest of the game-world art, it will stick out and ruin her immersion . . . If the player comes to a puzzle, figures out a perfectly reasonable solution to it, and that solution does not work, the player will again be reminded that she is 'only' playing a computer game.
>
> (Rouse 2001: 12–13)

For Rouse, the videogame player is positioned at the heart of the action – they effectively enter the gameworld. Importantly, this does not presuppose a first-person viewpoint, and as Newman (2002b) has noted, players may report first-hand experience of gameworlds presented in second-person view, as in *Gran Turismo*, third-person, as in *Sonic the Hedgehog*, or even via dynamically shifting viewpoints as in *Super Mario 64* or *Metal Gear Solid 2* (see Chapter 8 for more on viewpoint and player engagement and cf. Bates 2001). For this reason, videogames may be characterized by a sense of 'being there', rather than controlling, manipulating or perhaps even 'playing a game'. As Peter Main of Nintendo notes of *Super Mario World* (third-person viewpoint), 'Make no mistake, when these kids are playing Mario it's them up there on the screen' (*Serious Fun*, Channel 4, 1993). For many videogame designers, it is important to ensure that there is no explicit detachment and distance from the contents of the game and it is this desire that drives the creation

of ergonomic hardware and software interfaces, for example. The centrality of participation and the sense of 'being there' chimes with the stance of game theorists. 'Players concentrate wholly on the game – on the dice or the puck or the pawn; good gameplay . . . makes you forget yourself and the passage of time, not operating consciously but going with the flow' (Farley 2000). Moreover, this theme has been seized upon by proponents of interactive narratives who posit videogames as stories to be performed (see Murray 1997; Buse 1996; Laurel 1991; see also Chapter 6).

RULES, WINNING AND LOSING: VIDEOGAMES AS GAMES

The recent emergence of videogames as a legitimate object of study in media and cultural studies, in combination with the closer alliances between the videogames industry, Hollywood and the music industry (see Chapters 1 and 4), has had the, perhaps inevitable, consequence of underemphasizing their status as games to be played in favour of media texts to be read or consumed. In fact, as we will learn in Chapter 6, the appropriateness of approaches to the study of videogames is a matter of considerable discussion and debate in the field. However, as Aarseth (1997) and Frasca (2001a, 2001b) have noted, there is much to be gained by situating videogames within the broader critical context of game and play even though, as Frasca notes, 'ludology', the study of games and play, is comparatively poorly developed. While for some commentators, the characteristics of games are so diverse as to render any singular definition problematic (see Sutton-Smith 1997 on the variety and ambiguity of play activity), a body of literature has developed. Huizinga (1950), provides a useful point of entry.

The game is a voluntary activity, engagement with which represents an end in itself rather than operating as a means to an end; game play is its own reward and is clearly distinguished from ordinary life. Farley (2000) has critiqued this dislocation of gaming by pointing to the memorability of play either 'involuntarily' through physical injuries, or perhaps through the disgruntled recollection of a beaten player keen to improve or seek revenge. Central to Huizinga's conceptualization of the game and key to demarcating it as an experiential entity are the rules that bind, constrain and structure activity. Certainly, rules appear to be central to most definitions of games (e.g. Caillois 2001) and Huizinga's position is mirrored in the commonsense distinction between 'play' and 'game'. Where play is considered free and unfettered, the game is characterized by the constraints of rule systems. In attempting to chart a path through the enormous variety of 'game' activity both Piaget (1951) and Caillois (2001) seize on the variegated complexity of rules. Caillois distinguishes between 'paidea' and 'ludus' referring to games with simple and complex rules respectively. As such, skipping a rope (paidea) can be distinguished from more complex games such as bridge or football (ludus). Piaget's distinctions are similar in pointing to the comparative complexity of rules; however, his analysis differs in suggesting that certain types of game are, in fact, entirely free from rules. Importantly, according to Piaget, the shift from unbounded to rule-based gaming can be equated with childhood development. From birth to two years, children engage in kinaesthetic games of exercise in which they reach out and explore their surroundings; from two to seven, game play is characterized by symbolic role-play. Both types of game are, according to Piaget, free from rules and are distinguished from the games of children over the age of seven. For Piaget, these games with rules, such as football, are considered more 'adult' activities because they demand socialization. In commonsense terms, we can broadly equate Piaget's games with and without rules with 'play' and 'game' respectively.

Whilst Piaget's definitions are appealingly neat, Frasca encourages us to question the validity of the claim that games or even play is unbounded by rules. Daniel Vidart (1995) has noted that apparently unfettered play does, in fact, have strict rulesets. He gives the example of a child pretending to fly a plane by running around, arms outstretched. While following Piaget's classification, this might seem like unregulated, pre-socialized freedom, Vidart suggests that rules are at play here. The most obvious rule is that child has decided to behave like a pilot, and not a car driver, for example. As Frasca observes, while the rule is proposed and accepted by the player, and can be dropped at any moment, during play it is accepted like any other game rule. Rather than rules, Frasca suggests that play and games can be differentiated according to their outcome. Examining the work of André Lalande (1928), Frasca suggests that, unlike play in which there is no winner or loser, the result of games is victory or success. The distinction is also highlighted by Piaget who, after asking a group of children at play 'who won?' was greeted with mystified looks. The children did not understand the question. There is no winner or loser in play. In this way, the goal of the activity rather than the structure or constraints imposed upon players is key. Accepting Vidart's claim that rules are inherent in games and play, Frasca proposes a modification of Roger Caillois' (2001) terminology of 'paidea' and 'ludus' to describe this goal-oriented distinction. Paidea can be described as 'physical or mental activity which has no immediate useful objective, nor defined objective, and whose only reason to be is based in the pleasure experienced by the player' while ludus describes 'activity organized under a system of rules that defines a victory or a defeat, a gain or a loss' (Frasca 1999). Ludus, therefore, requires reference to an external suite of rules where paidea is understood and delimited by the player.

PAIDEA AND LUDUS IN VIDEOGAMES

It is tempting to think that videogames must be archetypal examples of ludus, however, Frasca suggests that both ludus and paidea can be observed. *SimCity* is an example of a paidea videogame. While the player can attempt to create an aesthetically beautiful city, or an efficient city, and while a session ends when resources are exhausted, it is not possible to win or lose. In his discussion of 'abdicating authorship', Doug Church (2000) has similarly noted that there is no winning or losing in videogames like *The Sims* beyond what the player understands those terms to mean. That is, the player imposes their own ludus rules upon the playground that *The Sims* offers. Moreover, Frasca suggests that ludus and paidea can be combined in specific games, and that the player themselves is able to switch between the two activities at will. When piloting an aircraft in *Microsoft Flight Simulator* or Nintendo's *Pilotwings 64* without a specific goal, the player is engaged in paidea. However, they can easily impose a ludus rule, such as to perform a particular manoeuvre without crashing. Thus, the notion of winning and losing is imposed. As Frasca notes, many videogames are specifically designed as environments for paidea and ludus:

> many flight simulators include different missions (particular goal-oriented activities where the player has to accomplish a certain task, such as bombing a city or landing under bad weather conditions). These *ludus* are hard-coded within the program: the program includes a ludus rule and it will tell the player if she has succeeded or not at the end of the session. However, the same mission-based simulator could also be used for *paidea*: the player could simply not follow the rule and would just play around with the airplane . . . It is the player and not the designer who decides how to use a toy, a game, or a videogame. The designer might suggest a set of rules, but the player always has the final decision.
>
> (Frasca 2001a: 13–14)

Thus, while videogames might appear (especially to the non-cognoscenti) to be restrictive experiences with many complicated, often impenetrable rules channelling the player into certain behaviours and responses, Frasca encourages us to consider videogames as worlds, or rather 'playgrounds' in which many different activities can be performed. Importantly, these playgrounds need not be restrictive but may be open and flexible and, while designers might suggest possibilities, it is ultimately players that decide which activities will be performed. For example, many games place the player in an initial situation from which they have to escape by traversing a landscape, environment, or in the case of *Luigi's Mansion*, a haunted house. However, the games do not tell the player how to conquer the game space, nor do they initially present any more than the barest of parameters for play. It is the job of the player to deduce (or even impose) rules through exploration, invention and imagination – reaching out into the world to test, evaluate and execute different approaches. Thus, while there may be one, and only one, way to capture each of the ghosts in *Luigi's Mansion*, it is left to the player to ascertain the appropriate approach. Importantly, even here, enacting the solution leaves considerable latitude for individual technique. Similarly, it is clear to the seasoned player that the majority of the various denizens populating Dinosaur Island, the setting for *Super Mario World*, follow deducible patterns of behaviour. However, neither this fact nor the specific behaviours are revealed to the player through printed instructions or on-screen tutorials. Rather, the player has to explore the gameworld, noting similarity and difference, identifying and matching patterns. For example, some Koopas will chase the player while others will resolutely patrol a limited patch of land, while each end-of-level opponent has its own unique set of qualities and weaknesses that need to be explored and exploited. The exploitation of rules may involve identifying tactics never intended by the game's designers. Perhaps the most infamous example is the *Asteroids* 'lurking' strategy. Rather than destroy all of the asteroids in the playfield and move on to the next level, experienced players learned that by leaving one floating through space, the level could be effectively suspended and they could wait for the arrival of the alien spaceships that earn far more points on destruction than mere rocks.

In observing children playing *Transport Tycoon*, Carsten Jessen (1995) has noted that working out the rules of a videogame constitutes a large part of the fascination and challenge and is a principal motivation for play. In fact, once the rules have been deduced and overcome, videogames may lose their appeal and new challenges may be sought, either through (purchasing) new games or the imposition of new ludus rules. Videogame play is principally concerned with exploration, testing out ideas and strategies. The demands made by videogames on the player's creativity and imagination are frequently overlooked in accounts of play that position games as stifling (see Grossman 2001; Dill and Dill 1998).

Deducing, collating, and working within or around a game's rulesets represents a large part of the pleasure of videogame play and further highlights the active, participatory role of the player. Livingstone (2002) has noted that the skills developed through their use of videogames during leisure time must be recognized as potentially crucial for ICT literacies. 'Far from representing an irrelevant or even problematic alternative to "serious" uses of computers, it might be argued that playing electronic games generates the kinds of skills and competencies that matter most for ICT use' (2002: 229). Livingstone continues to cite Johnson-Eilola's (1998) experience of his daughter's explanation of videogame play:

> To someone raised in an historical worldview – one valuing linearity, geneal-ogies, tradition, *rules* – Carolyn's explanations of the game sound haphazard, unplanned and immature. But to someone familiar with global informa-tion spaces such as the World Wide Web, games such as these provide environments for learning postmodernist approaches to communication and knowledge: navigation, constructive problem-solving, dynamic goal-construction.
>
> (Johnson-Eilola 1998: 188)

TYPES OF GAME

In addition to discussing the importance and complexity of rules, Caillois (2001) usefully identifies a variety of different types of game. Heavily influenced by Huizinga, Caillois proposes four distinct forms: *Agon* in which competition is dominant; *Alea* where chance and randomness are key; *Ilinx* in which pleasure is derived from movement; and *Mimicry* in which games are oriented around simulation, make-believe and role-play. Importantly, these characteristics are not mutually exclusive and so poker, for example, can be seen to combine elements of *Agon* and *Alea*. A consideration of even the most apparently simplistic videogames reveals the applicability and interplay of these various characteristics. In *Tetris*, for example, both competition (*Agon*) and chance (*Alea*) are evident. The randomness and unpredictability of the sequence of falling blocks ensures that *Tetris* cannot be simply 'learned' while competition can be between two (or more) players or between player and CPU (Central Processing Unit – essentially, a competition between the player and the game's 'simulation', see p. 25). It may be also that the element of competition is imposed by the player through their own ludus rules, for example, trying to maximize the number of four-line 'Tetris' scores. Farley (2000) has argued that all games are essentially agonistic and even those that require teamwork and co-operation do so only so that one team may vanquish another. Reconsidering Sega's coin-op *R360*, we note that in addition to *Agon*, the game clearly comprises *Mimicry* as the player adopts the role of fighter pilot and is spun around inside the motorized cockpit. Here, just as with 'bemani' dancing games, there is a quite literal kinaesthetic pleasure to play. Importantly, there is little randomness in these games with attacking fighter planes and dance steps following fixed patterns and the pleasures of both *R360* and bemani are to be found in the complex of movement, competition and role-play. These characteristics in combination with the modified, outcome-oriented ludus and paidea offer a means of differentiating (video)games from one another depending upon the relative balance of the elements. However, at the risk of singling out one characteristic above others, it is useful to further consider the nature of competition (*Agon*). Specifically, it is important to consider precisely with whom, or perhaps even what, the player is in competition. While we have alluded to multiplayer competition and even collaboration (see Chapter 9 for a fuller discussion), it is valuable to consider the videogame as a puzzle.

Puzzles in videogames present something of a problem as the term is slippery. Within industry and player parlance, puzzles usually refer to particular staged, set-piece episodes. As such, a puzzle may refer to the need to deduce the sequence in which a series of doors must be unlocked, or may require the use of a variety of collected items in a particular combination, at a particular time or even in a particular location. An over-reliance on puzzles is often cited as a criticism of games such as *Myst* or *The Seventh Guest* (see Juul 1999, 1998, for example).

Perhaps part of the reason is to be found in the disruptive effect of puzzles. As Farley (2000) has noted, for Huizinga (1950) part of the pleasure of the game is to be found in the presentation of an ordered world. This is in contrast with what Berger has called 'the casual and confused reign of everyday existence' (cited in Holquist 1968: 122). Indeed, for Turkle (1984), it is the seduction of this perfectible and manageable world of computers that should present cause for concern. According to Danesi (2002), part of the appeal of the puzzle arises from the disruption of this order, or more precisely, from the knowledge that order may be restored. Thus, the self-contained ordered world created within what Huizinga terms the 'magic circle' of the gameworld is disturbed (see also Jensen and Scott 1980 on the appropriateness of the physical setting or 'play landscape'). However, the reinstating of the equilibrium state may be seen to represent part of the challenge that Rouse (2001) and Crawford (1984) identify as central motivations for play. In this regard, we can observe that the structure of the videogame as puzzle appears quite similar to Todorov's (1977) narrative structure in which an initial equilibrium state is disrupted, recognized, tackled and ultimately resolved (see also Vogler 1998 and Chapter 6). Of concern to many videogame designers, however, is the puzzle with single victory state – a 'solution' (see Bates 2001, for example). To solve such a puzzle is to attain this state, and only this state. Examples of such single solution state puzzles include crosswords and jigsaws, for example, though the designation of videogames such as *Tetris* as 'puzzle games' is unsustainable given the absence of a solution (see Chapter 5 on the classic arcade game structure).

While Crawford is quick to dismiss the puzzle as defunct upon solution, Maroney (2001) notes that even puzzles with a single victory state may offer rich replay potential by introducing randomness into the initial state by shuffling a deck of cards, for example. In criticizing *The Seventh Guest* and *Myst*, what is condemned is the minimal scope for invention, experimentation or individuality. If what is denounced as 'puzzle-solving' is to be differentiated from 'gameplay' then it is perhaps in terms of the facilitation of strategy and tactical development and the operation of non-linearity. The issue that critics such as Rollings and Morris highlight is the orientation of the game around narrow, tightly defined puzzles that offer little latitude for creativity and limit the scope for individual solutions based around playing preferences, style or technique. Coupled with linear progression structures that demand the completion of one puzzle before the next can be attempted, the game can be seen as a series of episodes that require little more than the repetition or revelation of pre-ordained sequences of inputs or the marshalling of specific sequences of events. However, many videogames do not present such a structure and offer a variety of potential 'solutions'. Moreover, through the imposition of ludus rules, players can, themselves, decide to define the victory state. This may be related to the standard, external rules of the game (completing a section without firing, for example), or may be wholly unrelated (performing as many 'doughnuts' along the backstraight as possible). Indeed, players may choose to override the concept of victory states and indulge solely in paidea. Jensen and Scott (1980: 305–307) point to the absence of competition in the Hareskin Indians' 'keep-away' ball game and circle dance, and the Inuit modification of baseball into 'total community participation' with no winners or losers. In this way, players may modify games ostensibly designed with *Alea* in mind and remove the competitive element.

While it is usual to treat certain discrete sequences, episodes or elements within a videogame as puzzles (the location of a key or the placement of an object on a pressure pad to open a doorway), it is possible to view the videogame, as a whole, as a type of puzzle. Following Ted Friedman (1995), the act of playing a videogame can be conceived as an engagement with the apparatus of the simulation 'beneath' or 'behind' the game. The simulation that brings the gameworld and all its contents into being. The articulation of this dialogue varies between games. In some instances, the parameters of the simulation will be known to players at the outset of play, as, for example, in *Tetris*, where the extent and scope of the action is contained within a comparatively limited and inflexible, but explicit, ruleset. In such instances, the dialogue between the player and the simulation sees the evolution and origination of strategy and technique rather than the deduction, inference or exploration of rules. Here, the player contends with the element of *Alea* as they tackle the relentlessly descending blocks. However, as we have seen on p. 21 with *Transport Tycoon* and *Luigi's Mansion*, in other games, the exploration of not only the operation but also the boundaries of the simulation can be absolutely key.

To take to the track in *Gran Turismo* or square up to an opponent in *Tekken* is to explore the possibilities and potentialities offered. To play these games is to explore the limits of what is allowable within the context of the simulation. Friedman's (2002) analysis of the *Civilization* series is enlightening in this regard. For Friedman, the engagement with the videogame simulation as a puzzle demands that the player 'thinks like a computer'. While this phrase is potentially misleading as it conjures the imagery of 'cyberpunk' discourse (see Featherstone and Burrows 1995; Stone 1991; and see also Chapter 8) in its apparent technological determinism and anthropomorphism, it is nonetheless useful in capturing the sense in which the player is encouraged to consider the 'heart' of the game, looking beyond or behind the audio-visual presentation of the gameworld. It is useful also in stressing the exploratory, investigative nature of videogame play. As we shall explore throughout this book, the precise nature, balance and diversity of the investigation varies from game to game and may demand exploration and revelation of the various spaces generated by the simulation and that comprise and constitute the gameworld (see Chapter 7) or scrutiny of the artificial intelligence (AI) of that gameworld's inhabitants in order to develop strategies for success.

The consequences of a consideration centred on this engagement with 'the game' are far-reaching and we will learn in later chapters that there are implications for the ways in which we tackle not only the issue of the audience but also the representational system of the gameworld. However, any discussion of the videogame must be sensitive to the contexts in which the form is used and consumed. As such, and as we will learn in Chapter 6, even the definition of the videogame as oriented around the act of play and 'the player' is potentially problematic as we have already noted that videogames are not solely the preserve of the lone player and are often encountered socially with 'players' and 'non-players' sharing the experience and discussing and re-processing it through talk or reference to discussion boards and walkthroughs. Just as the act of play as encountering and deciphering the parameters of a simulation may present challenges to approaches to the study of video-games that offer primacy to the visual (see Newman 2001, for example), so too an understanding of the contexts of videogame use and the com-position of the audience potentially problematizes the centrality of the player in analyses.

VIDEOGAMES AND INTERACTIVITY

We have seen that the childish associations of 'game' and 'play' have led the videogames industry to seek a more respectable description of its activity. Its power as a contemporary marketing buzzword makes 'interactivity' an almost natural choice. Where novels, newspapers and cinema 'create' readers, the interactive audience is immediately empow-ered and placed at the centre of these new media experiences. However, the uncritical use of the term in a variety of contexts as qualitatively and experientially diverse as videogames and DVD scene access menus or, indeed, voting in such TV events as 'Great Britons' or 'Who Wants to Be a Millionaire?', has rendered it a fluid designation. For many theo-rists, 'interactivity' is such a nebulous and ideologically-charged term that replacements are sought. Aarseth (1997), for example, borrowing jargon from physics, prefers discussion of the 'ergodicity', or 'non-trivial' activity, that defines the cybertext. However, for Juul (1999: 21) interactivity need not be discarded: 'Computer games are interactive because the actions of the player play a part in determining the events in the game.' For some commentators (e.g. Frasca 2001a, 2001b; Murray 1997), it is this facility of the player through some manipulation exacted during their performance of play, such as the imposition or implemen-tation of a rule, for example, to affect a transformation on the game or 'text' that defines the interactivity of forms such as the videogame. For Crawford (1984), it is the interactivity of the game that differentiates it from the static puzzle. The game, or rather the simulation, responds to the effort and activity of the player. In this way, it is possible to differ-entiate videogame 'interactivity' from that offered by DVD menus wherein the ordering of material may be placed under the control of the viewer, but in which no transformative potential is afforded. Selecting one option over another allows the DVD viewer to re-sequence, timeshift or zip through the material (Ang 1996; Cubitt 1991) but not to alter the substantive content of sequences (note Livingstone 2002 on linearity and hypertextuality and Landow 1991 on the *lexia* and path structure (after Barthes 1974) of hypertexts and see Chapter 6).

The material revealed through menu selection is fixed unlike that in the videogame which, being oriented around a transformable, and importantly, responsive simulation, may dynamically adapt to the performance of the player (Ryan 2001; Murray 1997; Laurel 1991). However, as we will learn in Chapter 5, the discussion of whether videogames are 'interactive' or even 'ergodic' potentially overlooks an even more fundamental point. Videogames are highly complex, segmented arrangements of elements. Some of these elements may be seen to be highly 'interactive', requiring considerable player participation and responding to player action, while others, most obviously inter-level movie cut-scenes, appear to demand little or no direct player input or control, nor do they respond to attempts to exert influence. Yet, this is not to say that the player is not actively interrogating the material, exploring it for clues to aid forthcoming play or reading a presented narrative in order to make sense of past events or predict those yet to come. Videogame experience is, in fact, the product of a complex interplay of elements each demanding and facilitating different degrees and types of participation and activity.

SO, WHAT EXACTLY IS A VIDEOGAME?

Throughout this book, we will follow Frasca in using the term videogame in its broadest possible sense. That is, it will be used to describe:

> any forms of computer-based entertainment software, either textual or image-based, using any electronic platform such as personal computers or consoles and involving one or multiple players in a physical or networked environment.
>
> (Frasca 2001a: 4)

If a defining quality of the videogame is that it fosters the sense of first-hand participation in a gameworld generated by the computer, then we may be able to distinguish it from devices such as AIBO or Furby. In this way, videogame play may be understood as a form of 'embodied experience' (see Newman 2002b). However, while this participation is a defining feature of videogames, it is important to note that videogames do not offer a uniform experience of 'interactivity' and comprise sequences of high and low participation and differing modes of engagement. Following Caillois (2001) videogames offer combinations of chance, competition, role-play and kinaesthetic pleasures. Moreover, videogames can offer both paidea and ludus rules thereby allowing players to engage in goal-oriented or 'free play' activity. In this way, videogames are not merely to be viewed as restrictive rule systems and recognition is given to the necessity of exploration and deduction as well as the player's ability to ignore or even subvert a designer's intention. A player can develop tactics and strategy, perhaps exploiting weaknesses or flaws in the game, or they may even define their own games within the world made available, thus imposing their own ludus rules. Furthermore, the definition of videogames employed here recognizes that certain games – or certain sequences or modes within games – are designed as non-goal-orientated 'playgrounds'.

Videogame play can be understood as exploratory, open and free-roaming just as it can be puzzle-oriented and rule-based. Moreover, it is important to note also that the term 'player' is ambiguous as videogames are often experienced in groups with 'non-controlling' players, and are absorbed and understood within participatory cultures of talk both online and offline (Brooker 2002; Jenkins 1992). As such, concentration on just those clutching joypads reveals only part of the story and discussion of videogames as embodied experience can only account for the use of a portion of the audience. Importantly, accepting the problematic nature of delineating the audience, this definition of videogames does not require a technological demarcation; the definition is not concerned with screens, or other interface systems and we can comfortably discuss gaming experiences delivered through home consoles, coin-op cabinets or mobile devices, using graphical or non-graphical interfaces (see Livingstone 2002).

3

MANUFACTURING FUN
Platforms, development, publishing and creativity

VIDEOGAMES IN TRANSITION

We have already noted in Chapter 1 that videogames have existed, in one form or another, for over 40 years, and it will be clear to those who have even a cursory awareness of the products of the industry that videogames have changed significantly both formally and aesthetically. Indeed, it is hard to imagine that the primitive blips of light that formed the bats and ball of tennis-game *Pong* could be separated from the lavish, near-photorealism of *Metal Gear Solid 2*, *Halo*, or *Metroid Prime* by only 30 years, or that videogames could have so quickly become the global social, cultural and economic force they are today. In fact, far from being a smooth transition from the hobbyist's workshop to the development studio, the history of the videogame industry is one of turbulence, peak and trough. For a period during the 1980s, for example, under the weight of substandard product and consumer apathy, the global videogames industry fell into such a decline that it seemed unlikely it would ever recover. Since then, the situation has more than recovered and we have seen in Chapter 1 that videogames have (re)emerged as an extremely significant socio-cultural form and that the videogames industry is a major global concern.

However, intriguing though the history of videogames undoubtedly may be, this chapter does not seek to chart the events of the past 40 years. For those seeking such an overview, a number of excellent resources exist. Kent's (2001) *The Ultimate History of Video Games* is perhaps the most exhaustive overview of the industry and makes use of

extensive citations from influential designers, developers and commentators. The fortunes of individual companies are also documented, for
example see Sheff (1993) on Nintendo and its transition from playing
card manufacturer through to dominant player in the 1980s, Takahashi
(2002) on Microsoft, and Asakura (2000) and Nathan (1999) on Sony.
Online, *The Dot Eaters*, *History of Videogames* (see Herman *et al.* n.d.),
Classic Gamer Magazine, *Retrobase* and *Killer List of Videogames* (*KLOV*) are
among many sites that offer well-illustrated histories of aspects of the
history and development of the industry, videogame systems and experiences (see also *Videotopia* a travelling museum exhibition dedicated to
arcade videogames) while websites such as *Pong-Story* detail the gestation
and impact of specific titles.

Here, the discussion will focus on the ways in which videogames
have changed structurally, spatially and in terms of the interactive/
play potentialities they present to players with particular reference to
the relationship between design and technology. Furthermore, this
chapter will examine the transformations that have occurred in videogame development and, specifically, the procedural, methodological and
organizational changes that have necessarily accompanied the evolution
of videogames as 'artefacts'. Key among these changes has been the
emergence of increasingly formalized, managed development processes
and the coming of age of a videogames 'industry' in stark contrast to the
often ad hoc processes of 1970s' and 1980s' videogame production. The
diversity and complexity of the assets constituting the contemporary
videogame have seen the inevitable emergence of specialist teams and the
virtual death of the lone designer, responsible for all aspects of a game.
For many commentators, these large development teams with specialized, often ring-fenced areas of responsibility in combination with the
increasing dominance of frequently risk-averse financiers and publishers
required to fund and support large-scale projects, has led to a situation
in which the contemporary marketplace is largely bereft of the creativity
and invention associated with the heyday of development (see Kent
2001) and flooded with derivative products and sequels designed to safely
generate revenue.

VIDEOGAMES AND TECHNOLOGY

Videogames are inexorably linked with technology. Indeed, the advertising and marketing discourse that surrounds the launch of new
videogames consoles frequently makes greater reference to the implied
power of processors or the audio-visual capabilities of the system than
to the experiential potentialities of the games it hosts. As such, consoles

such as Sega's MegaDrive proudly display their technical credentials on their sleeve; '16-bit' is emblazoned on the top of the console in gold lettering in larger type even than the manufacturer's name. For Donald Norman (1998), such references to the technology 'under the hood' are wholly redundant and may be dismissed as nothing more than a smoke and mirrors marketing routine. These 'hygienic' features communicate little of the worth or usefulness of the products themselves or the experiences they may offer the player. Moreover, the widespread use of the jargon of computing may serve only to confuse or baffle the would-be player. It is also entirely true to note that, despite the claims of advertisers and marketers, technological development cannot be simply equated with superior games or gameplay. Such a technologically deterministic stance certainly cannot be sustained in light of the recent and growing interest in 'retro' that has seen 'vintage' or 'classic' games from the 1970s, 1980s, and even early 1990s, enjoy new leases of life as they are (re)discovered by publishers and players (see Chapter 10). However, technology is important and the development of gaming systems has had a profound impact on the form and structure of videogames. Thus, while avoiding discussions of bits, bytes, polygons-per-second and framerates, or the presentation of technological development as progression or advancement toward perfection, it is useful to note some of the ways that videogames have been transformed as new audio-visual, processing, storage and interface capabilities have arisen and designers have been offered new opportunities. The relationship between storage capacity, processing power and the production of videogame space is particularly revealing in this regard and presents a clear illustration of the interrelationship between technology, design and gameplay potentials.

SCROLLING, EXPLORATION AND MEMORY: PRODUCING AND STORING VIDEOGAME SPACES

As we shall learn in Chapter 7, space is key to videogames. Videogames not only offer worlds to inhabit and play in, but also frequently present puzzles and challenges that revolve around the occupation of space. From their earliest days, videogames such as Atari's *Asteroids* place the player in a hostile space, full of treacherous space debris and flying saucers and charge them with the task of surviving for as long as possible while gaining points by literally clearing the space and rendering it 'habitable'. Similarly, Taito's *Space Invaders* positions the player as the sole defence against alien attack. Again, hostile enemies occupy the game's space and the game may be seen as essentially a battle for territorial supremacy.

More recent games such as *Halo*, *Super Mario Sunshine* and *Luigi's Mansion* all present similarly spatial challenges where tasks encompass the reclamation of worlds, regions, locales, lands and buildings. However, while space remains a central theme, games such as *Super Mario Sunshine*, differ considerably from *Asteroids* or *Space Invaders*. Not only are the spaces presented in both *Asteroids* and *Space Invaders* representationally simplistic and rendered with considerable economy, but also both games take place within the confines of a single screen. Everything within these worlds is visible at all times. The screen is a literal spatial boundary in *Asteroids*. Indeed, attempting to fly 'off' or 'out of' the screen simply results in the player's ship reappearing at the edge of the screen on the opposite side; fly off to the right, reappear at the left; fly out of the top of the screen, reappear at the bottom. The screen is inescapable and, importantly, immovable. It presents a single, fixed view of the world that it contains and the extent of that world defined by the screen. This is in contrast to games such as *Halo* and *Super Mario Sunshine* where the gameworlds are far larger than can be displayed on a single screen and must be traversed and, most importantly, explored. In fact, it is this exploration, this journey through space, that represents at least part of the pleasure of these games. It is clear that such exploration is, at best, limited in single-screen gameworlds.

Here, then, we note a significant shift in the design, implementation and pleasure of videogames that arises from a lifting of technical constraints. The early, single-screen gameworlds of the 1970s must be seen as the products of the technical restrictions placed on designers by the limited graphical capabilities of the hardware they worked with. With developments in visual processing technologies came the possibility of creating physically larger, more expansive, and more complicated, game-worlds. At first, games were constructed of many interconnected, yet still resolutely single, screens that comprised a larger spatial whole. Then, with scrolling (the movement of the 'background' or gameworld to create the illusion of movement) first in two-dimensional and later three-dimensional space, it became feasible to produce expansive, contiguous spaces. While text-only games such as MUDs (Multi-User Domains) had virtually presented complex worlds through which players could travel, with scrolling, these worlds could be rendered visually and in real-time. The technological developments in visual processing that repositioned the screen from 'boundary' to 'window on a (larger) world', have had profound effects on game design and new types of gameplay experience have been facilitated. Where single-screen games had demanded that players protect spaces under threat or more simply had presented their space as a frame within which to engage in combat, multiple-screen and scrolling games are frequently oriented around the thorough exploration and scrutiny of the spaces they produce. Thus, space is not only a container for the action, but is a central constituent of the game. It is part of the challenge, the solution and even the reward of the game, and must be traversed, explored, utilized, mastered and perhaps even conquered.

The production of space in videogames is tightly linked to the capability of any given system and videogame spaces bear the fingerprint of the technical limitations or freedoms offered to designers. In addition to scrolling and the shift away from the containment of the screen, in the mid-1990s consoles such as PlayStation, Nintendo 64 and Saturn brought the ability to generate real-time three-dimensional worlds and videogame designers seized the opportunity to offer yet more complex spaces to investigate, play and puzzle in. Taking advantage of powerful visual processing tools and large amounts of memory space to store virtual geographical data, Nintendo's *Super Mario 64* and *Legend of Zelda: Ocarina of Time* are but two examples of the sprawling worlds conjured by designers within which players could immerse themselves. Navigation and familiarity with the spaces and places within the game are vital skills without which prowess with a sword, for example, are inconsequential. Not only do such spaces present still richer potential for player exploration, but also they heighten the kinaesthetic pleasures of play with performance moving seamlessly through extensive, topographically varied, worlds.

COMPLEXITY AND DIVERSITY

One consequence of the heightened spatial and experiential diversity of videogames has been a complication of the processes of production. The range of assets – and it follows, specialist individuals and teams – required to produce a game such as *Metal Gear Solid 2*, has necessitated a shift in development practice. Most notably, the complexity and diversity of such games demands a more formalized, managed development strategy and effectively prohibits the 'one man band' operations commonplace in the 1970s and 1980s. The composition of the modern developer, as we shall see, borrows in part from the film industry and in part from the non-entertainment software industry (see Cringely 1996).

The impact of this transformation of development procedures may be felt in the marketplace. Certainly, also, it affects the prospects of individuals keen to enter the industry. Where a range of skills and the ability to turn one's hand to coding, level design, character design and even audio and music composition was not merely desirable but essential in previous decades, scrutiny of employment websites and journals such as *Computer Trade Weekly*, *Edge* and *gamasutra.com*, reveals that contemporary developers seek honed specialism in often narrowly defined areas. As such, those versed in 'physics engine programming' (the creation of parts of the computer program that define and simulate the existence and effect of factors such gravity and inertia), need no expertise in graphic or level design, for example, and perhaps even no experience of the videogames industry. The complexity of modern consoles and the player expectation for videogames all but rules out the lone designer/ programmer. Yet, from the earliest days of *Spacewar*, *Computer Space* and *Pong*, to the heyday of the 8-bit home computers and even beyond, the lone game developer reigned supreme.

During the 1970s and 1980s, it was entirely usual to find multi-million-selling videogames created by individuals operating from the makeshift surroundings of bedrooms or converted garages. Given this change, then, it is worth exploring just what is meant by 'creating' a videogame in this sense. These solo game developers could be responsible for not only the design of the game (assuming that they were dealing with an original title rather than a conversion), but also its coding, its artwork including backgrounds and character animation, level design, and perhaps even the music and sound effects. It is clear that such a diversity of tasks demanded individuals with diverse skillsets encompassing programming and art, level design and puzzle creation. If ever there was an industry where multiskilling was necessary, this was surely it.

At least one reason why the solo developer was so prevalent can be attributed to the technology they were working with. Early game systems were simple and, certainly by today's standards, woefully underpowered. While comparisons between the technical capabilities and storage capacities of modern gaming systems like Xbox, PS2, GameCube and 1980s' home computers such as the Sinclair Spectrum or Commodore 64 are misleading, it is worth noting that, while discussion of today's systems centres on how many millions of polygons a game draws, it is only 20 years ago that players and developers got excited about games being presented in colour, and being able to put eight moving objects on the screen simultaneously was a considerable technical achievement. While the technical constraint of early gaming technology might seem oppressive, it can be argued that it is precisely these restrictions that gave rise to some of the most enjoyable and playable games ever designed. Quite simply, it was not possible for the designer to wow the player with eye (or ear) candy. The sumptuous movie sequences players have become accustomed to on PlayStation or Xbox were just not possible. If you wanted to watch a movie, you had to go to the cinema. Games were about playing, so designers had to concentrate on making the experience of interaction as compelling as possible. *Pac-Man*, *Space Invaders*, *Pong*, *Defender*, *Paradroid*, *Monty on the Run*, *Manic Miner* are not audio-visual spectacles per se, although they retain an almost naive charm that owes much to nostalgia and retro-chic, but they remain excellent games, because, among other things, they are well balanced and perfectly paced.

While it is generally true that game development was a solitary practice, it is not correct to suggest that all games created in the 1980s were the result of endeavour by individuals alone. The most likely reason for teams emerging would be that an individual, while gifted in the areas of game design and coding, possessed no artistic abilities either in graphic design or musical composition. Consequently, a team might comprise the designer/coder aided by an artist producing the in-game graphics, and a musician. Very often the musician would be a freelancer and very often a coder themselves. The videogame musician's coding skills were extremely important. With the advent of consoles such as the PlayStation, musicians have been able to contribute to games by providing music in standard audio formats. With all the equipment and processing of the modern recording studio at their disposal, modern game musicians can present mastered music ready for pressing to CD. In the 1970s and 1980s, consoles and home computers had no CD playback (in fact, CDs did not even exist for most of this period!) and in-game music was performed on the game device's built-in sound chip. As a result, the music was stored as part of the program data and the musician would often write their own driver to turn the code into sound. Such were the limitations of the hardware, this meant that, in the case if the Commodore 64, for example, the musician had to not only write a memorable tune, but also write it using no more than three notes at a time, and write it in machine code, and put it all together in just a few kilobytes.

While a group of three people working on a project could be conceived as a development team, as Rollings and Morris (2000: 165) note, communication between designer/coder, artist and musician could be quite minimal and the areas of work kept quite discrete. In reality, despite the contributions and assistance of other specialists where necessary, game development remained an individualistic process. The simplicity of the gaming systems they were designing for, coupled with minimal team working, meant that it was quite feasible for an individual to have the overall vision for a game in their head. Working alone, it was simply not necessary to formally write up the game or produce development documentation. Moreover, it was not unusual for the game development cycle to be as short as just a few weeks. For example, Matthew Smith, creator of the seminal *Manic Miner* for the Spectrum, put the game together in just eight weeks. Twenty levels of platform action, with a novel aesthetic, beautifully drawn graphics and a pixel-perfect collision detection routine, had been created by an unemployed 17-year-old in his bedroom in just eight weeks ('The making of Manic Miner', *Edge* 2001: 95). It is hardly surprising that time was better spent designing and implementing rather than planning and documenting.

The programmable home computer gave rise to an industry that was characterized by small, independent developers. Compared with the contemporary industry with its big budget, tightly managed development processes and diverse range of assets requiring specialized teams, the technical limitations of early home computers necessarily privileged the game concept and play mechanic. In essence, a developer could not hide a bad game behind an audio-visual veneer, or a licensed franchise as Atari found out with *ET* and *Pac-Man*. While it is tempting to consider the constant march of technical development as inevitably leading to videogaming perfection, for many, the severe technical restrictions of early gaming platforms gave rise to the industry's heyday, pushing creative game designers to the fore. The current interest in 'retro gaming' and the vibrant trade in second-hand computers, consoles and software is a powerful endorsement of the enduring pleasures of early games and is, perhaps, evidence of a longing for the reinstatement of the balance between creativity and imagination and technical proficiency in implementation.

Manic Miner's development highlights other issues. The explosion of comparatively cheap, programmable home computers like the Spectrum and C64 had encouraged a new generation of programmers and game designers, and they were young. There is a curious circularity about the game industry in the 1980s during the home computer boom, which saw young boys in bedrooms creating games to be played by young boys in bedrooms. The popular image of gaming as an adolescent pursuit must, at least in part, be attributable to this arrangement. Moreover, it must surely be possible to trace its continued status as something other than a serious media industry to these days of the enthusiastic amateur.

THE MODERN DEVELOPMENT STUDIO

The sheer amount of money, length of time and number of personnel involved dictated that the haphazard, undocumented, and in many cases, unplanned development that gave rise to many classic videogames, had to change. Consequently, industry discussion forums and journals frequently discuss team management and project planning alongside gameplay design and balance (see Davies 2001, for example, on development methodologies).

It is tempting to think that videogame development must have changed immeasurably since these early days and that what was once a cottage industry finding its feet has established working practices and defined roles and responsibilities. Certainly, budgets for game development have grown, as has the development cycle and it is not uncommon for games to be in development for many years. Gaming systems have become more technically complex and offer potentials far beyond the imagination of the bedroom coders of 20 years ago. Quite simply, perfecting artificial intelligence routines, designing complex, sprawling levels, inventing characters and drawing animation cycles, programming physics engines to simulate gravity or the precise reactions of a car as it collides into a wall, scripting, directing and rendering introductory movies and cut-scenes, are beyond the ability of the individual.

While technology was simple, the scope of projects meant that they were manageable to the individual. Today's games, and today's game systems, require development teams and, most importantly, more formal development management and methodologies. As games increase in complexity and scope, even the development teams grow in size. For example, the development team for Konami's *Metal Gear Solid* comprised some 15 members. For the PlayStation 2 sequel, this had grown to over 70 ('The making of MGS2', FunTV 2002) with concomitant spiralling budgets. Kojima states that the budget for MGS2, for example, was approximately $10 million (Keighley 2001: 5).

DIVISIONS AND ROLES

Given this increasing complexity, it is perhaps inevitable that a variety of discrete roles and divisions have begun to emerge. Rollings and Morris (2000: 180) identify some of the key areas of specialization (Table 3.1).

It is important to note that these roles and divisions are by no means fixed and may not be found in every developer. Moreover, these roles do not necessarily equate with specific jobs within the industry and it is quite possible, for example, that an individual may perform more than one role. Furthermore, in all but the smallest developers, it is likely that a number of projects will be in development simultaneously. In business terms, this is something of a necessity given the long development cycle of the modern title, so it makes commercial sense to have projects coming to fruition at different times.

Table 3.1 Videogame development roles and divisions

Management and design	Game designer
	Level designer
	Software planner
	Lead architect
	Producer (Project manager)
Programming	Lead programmer
	Programmer
Art	Lead artist
	Artist
Music and miscellaneous	Musician
	Sound effects technician
	Motion capture technician
Quality assurance	QA lead
	QA technician
	Playtester

Source: Modified from Rollings and Morris 2000: 180.

One important implication of having multiple projects in development is that resources and personnel can be shared. Many developers arrange themselves so as to make use of a pool of talent deployed across a range of games rather than dedicate individuals to single titles. For example, an artist, or team of artists, may create graphics for more than one title and it is highly likely that an in-house composer will produce music for each title in development. Clearly, with such an arrangement, management and direction are critical in order that projects maintain focus and coherence. For this reason, different divisions are headed by a 'lead' whose responsibility it is to ensure that the various elements of a team's work retain consistency and contribute to the overall vision of the game.

The emergence of the areas of 'Management and design' and 'Quality assurance' are perhaps the two most telling events in the recent history of the industry. They are a salient representation of the nature and scope of the impact on videogame development brought about by the technological and marketplace changes seen in recent years.

MANAGEMENT AND DESIGN

The emergence of a formal management stratum has been a crucial factor in videogame development. Where pioneers like Matthew Smith worked alone, modern development is conducted in studios and projects are meticulously planned. The degree of co-ordination and the proliferation of employees required for such projects should give an indication of the level of investment that now takes place. Managing the sheer number of personnel involved in the creation of a game and marshalling the creation of the diverse assets required has become an important element of the process. Moreover, ensuring that every member of the team is aware of the overall aim of the game and how their contribution fits in is essential. Game designer Hideo Kojima ensured participation and inclusion during the development of *Metal Gear Solid 2* by encouraging every team member to keep 'idea notebooks'. As assistant director Yoshikazu Matsuhana notes, Kojima perused the notebooks each evening. Once programming tests were conducted, and assuming the feature added to the pleasure of the gameplay, it would be adopted. A number of trademark features of *Metal Gear Solid 2*, such as the ability to look around walls, were conceived in this way (Matsuhana in 'The making of MGS2', FunTV 2002).

The change from a form of 'individual authorship' to an industry of game development is, perhaps, analogous to creative scenarios in other media. Pedersen certainly seems to think that this is the case in respect of design:

> The game designer is the visionary, somewhat like a book's author. This person has outlined the scope and description of the product with sufficient detail so that others can understand and develop the product. Just as the book author sees his creation develop differently when made into a film, the game designer needs to accept and solicit modifications from the team members, the publisher and the public during the development process.
>
> (2001: 1)

The initial responsibility of the game designer is to produce the equivalent of a film treatment outlining the distinctive features of the game. As with a writer approaching a film producer, the concept documentation is primarily created to pitch to potential publishers (assuming the developer does not publish its own titles, see below) in order to get the green light to proceed. The concept document is usually written in conjunction with the producer, lead programmer, lead artist and the marketing department. Because it is principally conceived to support the pitch to financiers, concept documentation is generally concise and punchy, discussing the functionality of the game and its potential position in the marketplace. Consequently, once a project has been green lighted, concept documentation is of little use during game development.

Whereas concept documentation has little impact on the actual design process, the 'design document' created by the game designer is of primary importance. In other parts of the software industry, the design document is known as the 'functional specification' but, in reality, it is more than this. Design documentation describes the gameplay, the types of level or scenario to be encountered, and the variety of objects, weapons and enemies. Depending on the type of game, the design document is the place where the storyline, or the backstory that contextualizes the game, is mapped out. This element of the design document is often known as the 'story bible'. However, while part of the purpose of the design document is to provide a reference for all members of the development team, as Rouse (2001: 300) notes, it can also prove extremely useful to future development teams converting the game to other platforms. Bartlett (2000: 1–2) has observed that the role of the game designer is perhaps the most blurred and indistinct of all in game development and different designers have different sets of skills and perform different tasks during development with some involved in level and even character design. Such 'multiskilling' is typical in contemporary industry, yet particularly pertinent here because it indicates that the design process is not a 'one-off' but an ongoing process which will facilitate updating and recycling of existing product and even the creation of new games.

While it is generally true that the game designer need not be a top-flight programmer amid his/her multiskilling, a knowledge of current technology is certainly advisable if for no other reason than to gauge the feasibility of projects and designs. Such feasibility will derive from a knowledge of the market and a sense of the competition in the marketplace. The essential function of the management and design teams are thus twofold. First, they must ensure the coherence of the creative vision: what will and will not work. Second, they must track the progress of the development process in relation to launch dates and other interim milestones which relate to the marketplace in general, but specifically to the activities of competitors. The increased complexity of videogames, then, has been intimately related to the size and diversity of a production company's assets, its production teams and resources and the same factors which accrue to its competition in the marketplace. The need for tighter management of the production process and more formalized, accountable development methodologies reflects the considerable financial investment in projects such as *Metal Gear Solid 2*.

QUALITY ASSURANCE

The complexity of contemporary videogames, both in terms of their construction from elements created by different teams, and the breadth of experiential freedom they offer players, means that the importance of 'quality assurance' is difficult to overstate. Indeed, some argue that even existing levels of quality assurance are simply too low. For Peter Molyneux, the failure of the videogames industry to sufficiently employ playtesting and other quality assurance activities, is significant:

> Everyone says, 'Well, why aren't games better – why aren't there more really good games?' And I think that the answer is that what this industry doesn't do, amazingly, is play the games it makes. We create a game, we ask the teams to work all the hours God sends, and we don't give them time to play the game. That's really what makes the difference – sitting down and playing for hours and hours.
>
> (Molyneux in Rouse 2001: 472)

If the videogames industry is to respond to player expectations by creating more flexible games with greater open-endeness, non-linearity, and a focus on allowing increased player freedom of strategic or tactical exploration, then quality assurance is essential.

Harvey Smith, lead designer on *Deus Ex* and project director of *Deus Ex 2* illustrates the ways in which increasingly sophisticated, flexible games that offer players scope for exploration and experimentation can cause unanticipated problems, the identification and solution of which must be factored into the development process. In most cases, 'emergent' gameplay is seen as positive and desirable. Game designers present players with adaptable, open scenarios within which they can test different strategies and techniques and gameplay 'emerges' as a result of the player's engagement with, and exploration of, the situation and their capabilities. However, emergent behaviours are not always compatible with the game designer's intentions, as the following simple example demonstrates:

> When we did succeed in implementing gameplay in ways that allowed the player a greater degree of freedom, players did things that surprised us. For instance, some clever players figured out that they could attach a proximity mine to the wall and hop up onto it (because it was physically solid and therefore became a small ledge, essentially). So then these players would attach a second mine a bit higher, hop up onto the prox mine, reach back and remove the first proximity mine, replace it higher on the wall, hop up

one step higher, and then repeat, thus climbing any wall in the game, escaping our carefully predefined boundaries. This is obviously a case where – had we known beforehand about the ways in which these tools could be exploited – we might have capped the height or something.

(Smith, H. 2001: 2)

In light of this example, it is useful to distinguish between two distinct dimensions of videogame QA: 'debugging' (concerned with potential technical faults) and 'playtesting' (concerned with the target market's enjoyment). Where the former is a matter of ensuring the delivery of code that does not crash or freeze during play, the latter is interested principally in the way the game actually plays. Debugging is an essential part of the programming process. Software bugs may cause memory leaks that hog system resource and create bottlenecks while others may cause more immediate and noticeable problems; graphics may not draw properly and may even cause the game to crash.

Where QA technicians are concerned with bug fixing and the integrity and functionality of the code, the playtester is interested in the way the game behaves and what it feels like to play. For Rouse (2000: 473), playtesting is essentially 'bug fixing the game design'. Issues such as balance are very much the focus of playtesting. Cory Nelson, testing manager at Interplay Studios highlights the issue: 'If you spend two hours getting past a monster and only earn two points, there's something wrong' (Nelson in Crosby 2002c: 2; see also Crosby 2002a, 2002b). Importantly, and frequently overlooked, part of the role of the tester is to ascertain whether or not the game is actually 'enjoyable' to play. So, unlike technical bug fixing that can be more easily conducted using quantitative measures, playtesting relies heavily on subjective opinion and qualitative indicators. This raises a number of important issues. First, it is important to select the right testers. Too often, playtesting is conducted by members of the development team who are too close to the project to offer the detached, critical interrogation required. Similarly, friends, colleagues, and marketing departments all have their own agendas or are hamstrung in different ways and unable to offer their true opinion. Second, while most players can judge a good game or level from a bad one, far fewer can explore precisely why this might be, and fewer still can explain it in terms that are useful to the development team.

In addition to in-house playtesting, many game developers are following other areas of the software industry and conducting larger-scale 'public beta tests'. The public beta test involves releasing a nearly-completed version of the game to the game-playing public (usually with some functional limitation, perhaps timing-out or missing certain critical features such as the inability to save). As Rollings and Morris (2000: 186) note, distribution can be open, where beta versions are distributed via the web or on magazine cover-discs as id Software's *Quake* and *Quake II*, or may be limited to specific groups of testers as Origin's *Ultima Online*. In a somewhat ironic twist, it is increasingly common to find that beta testers are required to pay a small fee in order to take part in the testing process.

VIDEOGAME PLATFORMS

As modern videogame players are well used to retail shelves heaving under the weight of many hundreds of games for each games system, it is easy to forget that early systems were designed to play just one game only. Machines were 'hardwired' for particular games and could not be reprogrammed. A home *Pong* console was designed for the singular purpose of playing *Pong* and if the player desired a new game or simply tired of *Pong*, the machine became effectively redundant. In fact, though they did not invent the technology, it was not until Atari released its VCS (Video Computer System) in 1977 that the model of the multi-purpose console with interchangeable cartridges became dominant and the single-game system was abandoned.

The implications of the move to a multi-purpose system are manifold. Most importantly, where manufacturers of single-game systems were inevitably caught in the ebb and flow of fashion and the predilection for particular games, multiple-game systems offered potential longevity and the ability to reinvent themselves as fashion and fad dictated with a simple cartridge swap. For players, the attraction is at least partly to be found in the neatness of a single system with a library of games rather than having to own, set up and maintain a discrete system per game. In addition, there are potential financial savings as peripherals such as controllers become one-off purchases rather than comprising part of the cost of each dedicated, single-game system.

This stability of the base system represents a crucial turning point in the history of videogame development in terms both of the types of games and the nature of development. For manufacturers, the move from single-purpose to multi-purpose videogame systems, or 'platforms', makes possible an entirely different business model in which software is the principal revenue stream. As Sheff (1993), among others, has noted, videogames hardware – the consoles – are usually sold at, or near, cost and even at a loss, the object of the exercise being to get as many as possible installed in homes. The principle is broadly the same as that which applies for the installation of satellite dishes, set-top boxes or cable networking for the viewing of satellite, digital or cable TV channels. Whereas TV companies rely on subscription fees to counter any loss or lack of profitability, videogames manufacturers operate on the understanding that players are unlikely to purchase just one game for their system, and with systems offering a lifespan of many years (nearly a decade in the case of PSone), software sales are a potentially high volume business.

This business model clearly makes software development extremely important. First, it means that hardware manufacturers wishing to reap the financial rewards of high volume software sales themselves, frequently turn their attentions to development or publishing activities (see Microsoft, Nintendo, Sega and Sony, for example). Second, to ensure the popularity and longevity of their platforms, it is essential to ensure the continued support of 'third-party' developers. That is, those developers who are not part of or affiliated with the hardware manufacturer themselves. Such relations ensure a supply of both the most popular games, or the 'best' version of a game available on multiple platforms. These factors, in turn, have an impact on design. In the case of some platforms, the creation of additional levels, stages or sequences might be required or an increased level of elegance, efficiency and fluidity of the program might be instituted.

Yet, the key strength of a platform – its flexibility in relation to diverse software – is thereby also its weakness. It can play host to a variety of different games and even different game types, yet it is a compromised design that is not necessarily well-suited – and certainly not optimized or tailored – to any of them. The ability to utilize the power of modern PCs to run videogames for a variety of coin-op and home consoles, has heightened this issue still further as even the distinctiveness of different platforms is lost. Where Sega MegaDrive, Super Nintendo Entertainment System, *Defender* and *R-Type* coin-ops may have presented quite different input controllers, running under emulation on a PC, Mac or even PlayStation 2 or Xbox, each of these games and systems (among many others) is homogenized and makes use of a new suite of 'standard' controls. Where once *Defender* required the use of an unwieldy array of buttons and switches, on Xbox the standard joypad is deployed and so, while the game is reproduced in all its bona fide audiovisual splendour, the experiential business of controlling the spacecraft is adequately rather than authentically translated.

It is possible to argue that the adoption of the platform model stifles the creativity of games designers in forcing them to utilize standard hardware devices and software tools and that games may be, first and foremost, designed to suit the capabilities and strengths of the system rather than game designs preceding and dictating technical implementation. Free Radical's *Timesplitters* is a case in point. Keen to ensure a release date that coincided with the launch of the European PlayStation 2 console, the design ethos consciously sacrificed audio-visual extravagance in favour of efficient code, the minimization of development time and the delivery of fluid gameplay. In practice, this meant, among other things, trading visual complexity for speed of movement through the gameworld. At least part of the object was to maximize the productivity of the development team as they encountered the new platform.

This raises another important and often overlooked issue. The modern videogames industry, as we shall learn below, plays host to a great number of sequels and series. Aside from the marketing appeal of such 'cash-cow' franchises as *Tomb Raider*, *Street Fighter II et al.*, it is often the case that development is already geared to sequel generation. It has been suggested that this is the case in relation to the flexibility built into the design process. However, design teams often either explicitly treat the first incarnation of a series as a technology test-bed, or simply learn from their experience and mistakes and are capable of extracting greater performance from the hardware in subsequent versions of the series. Even though this may seem, in some ways, an ideal means to reproduce capital, the forces of marketing and development may be in tension with one another. With the benefit of experience, development teams may be able to design and implement experientially and technically more fulfilling and proficient games; but publishing and marketing pressures may lead to truncated development schedules that limit creativity in both design and implementation. Motivated by a desire to deliver products at strategically significant points, whether seasonal or in relation to the release dates of other games perhaps on other platforms, sequels such as *Devil May Cry 2* often appear pale reworkings of their forebears, hinting only in places at the invention and technical competence of the developers. In this way, it may be possible to see publishers and marketers as responsible not only for the existence of sequels and series, but also, in their shaping of the design process, responsible for the games' frequent mediocrity. Ultimately, this may have financial and design implications for the games industry. Comparisons have been made with the mid-1980s crash in the industry wherein consumers, disillusioned by a marketplace overrun with low-quality products, turned their backs on videogames en masse (see Demaria and Wilson 2002; Kent 2001; Sheff 1993).

FINANCE, PUBLISHING AND RISK

In order to maintain a market for its product, the games industry must continue to satisfy consumers through intelligent and sensitive implementation of design. Within player, reviewer and developer communities, perhaps the most frequent criticism of the contemporary videogame marketplace concerns the lack of original content, derivative gameplay, and the proliferation of licensed products (movie tie-ins, for example) and, particularly, sequels. As Nick Gibson of Durlacher notes, 'If you look at EA [Electronic Arts, US videogame publisher] over 50 per cent of its turnover comes from sequels, franchises and licenses that it can reproduce on an annual basis' ('Raising the stakes in the funding lottery', *Edge* 2002: 7). Certainly, EA's catalogue includes a variety of sports titles, such as the officially-licensed *FIFA* series whose main overhaul each year is mainly graphical and statistical, merely ensuring the correct player names and kit colours, and furnishing the re-release with the appropriate year suffix.

Given the sea-changes in the technical potentialities of videogames technology that have accompanied each 'generation' of PC or console, the charge that games are derivative and lack innovation is, in one sense, surprising. However, as Celia Pearce notes, while technological advances have rendered strictly unnecessary such unsophisticated mechanics as 'fight or flight', they nevertheless persist. While for Pearce, this reflects their continued popularity, Castle takes a rather different stance, citing the conservatism of publishers and its impact on player expectations: 'We're at a point', says Castle:

> where we can do something else, but we won't. We won't because we're afraid, because our market has become addicted, if you will, to this particular methodology. They want to try something a little new, but not a *lot* of something new. And people don't want to put up millions and millions of dollars to find out if there's another audience out there. And they don't even know how to educate them if they were.
>
> (Castle, interviewed in Pearce 2002)

The increased costs of development coupled with the need for publishers to generate consistent revenue streams to satisfy their investors has led not only to the formalization of development processes as we have seen, but also to a finance and support system that, broadly, is risk-averse and cautious.

Certainly, observations on the stifling caution of the industry need not suggest that all contemporary videogame design is simply formulaic and bows to the pressures of the market while sacrificing artistic or creative vision. Titles such as *Rez*, *Frequency* and *Parappa the Rapper* are just some of the many recent games that have offered genuine innovation. Likewise, the temptation to create nostalgic visions of the past as places in which every new game was ground-breaking and every developer a pioneer, must be resisted. What is important to note, however, is that, as a direct consequence of the investment required for the development of a contemporary videogame, the contemporary industry is markedly different from even ten years ago. With publishers increasingly wishing to see games at a later stage of development before offering backing in order to guard against the possibility of investing in projects that will not come to fruition, it follows that larger, more financially stable, developers that can use the profits from previous projects to support new developments, succeed at the expense of smaller operations. Being able to fund a project further into its development cycle places a developer at a distinct advantage when pitching to a risk-averse publisher. Supporting a project on the basis of an early concept, with (perhaps) an uncertain amount of development work still to be undertaken, represents a risk to the publisher. Progressing beyond this point into preproduction and the creation of workable prototypes or demos, for example, offers publishers evidence of a more tangible product in which to invest.

The catch-22 situation is one in which publishers will not fund development beyond the concept stage, yet in which development beyond concept stage cannot proceed without the financial support of a publisher. One response to this predicament has been the establishment of organizations such as Fund4Games, IFinance, WiseMonkey and Start Games that seek to assist developers by financing the early stages of game design. Indeed, in the case of Capital Entertainment Group, projects are hand-selected, funded to completion, and creatively and managerially supported. Principally concerned with helping the production of proof-of-concept documentation and playable demos to pitch to publishers, Start Games, for example, seeks not only to finance development, but also to support and nurture creativity by allowing developers the opportunity to present innovative ideas to publishers.

The impact of the emergence of the videogames industry has been profound. From the formalization of development methodologies to the conservatism of publishers and the marginalization of the small developer, let alone lone coder, the consequences of both technological and institutional change are considerable. However, the videogames industry has witnessed a history of increasing complexity and accountability in the design process. Far from entailing the onward, steady march of videogames' progress, this has, in fact, inculcated degrees of inertia.

The variety and complexity of assets required for contemporary games to meet consumer expectations has been coupled with the enormity of the sums required for investment in projects of this nature. The demands of publishers and financiers that revenue streams be not just significant but dependable and constant has given rise to a large videogames industry. The constraints of the contemporary design process, the demands on the marketing of titles in a highly competitive marketplace and the double-edged dominance of games platforms has led to a situation in which some see creativity and innovation being stifled. One response among players – the audience for videogames – has been involvement in the 'retrogaming' scene, offering an opportunity to recapture some of the innovation and invention deemed lost to the interests of the current marketplace.

In the next chapter, we will examine the changing nature of the audience for videogames and the many ways of defining, delimiting and studying it. In addition, and anticipating the more detailed discussion of social gaming and retrogaming in Chapters 9 and 10, we will note the ways in which players signal their position within the wider audience and, in particular, those who consider themselves members of a clique of 'hardcore' gamers, opposed to what they consider to be the derivative products of a contemporary industry sacrificing creativity for high volume sales.

4

VIDEOGAME PLAYERS

Who plays, for how long and what it's doing to them

THE CONTINUING MYTH OF THE VIDEOGAME AUDIENCE

While it has been noted that videogame play is an extremely popular activity, precisely who is playing is less well known. Throughout the course of this chapter, we will note a variety of ways in which the videogame audience is, and has been, understood and delineated by marketers, researchers and players themselves and the motivations that underpin these conceptualizations. As such, we will see that the 'audience' is a slippery entity that is variously considered as a group of market researched users, a group to be targeted and sold products, and a group not defined through empirical research but rather 'read' or 'implied' from the text of the videogame.

In 1991, in an important early study of videogames Eugene Provenzo noted that the core audience for games was adolescent boys. More recently, Shuker (1995) cites Nintendo's own demographics surveys which show that 36 per cent of their users are boys aged 8–11 with some 34.5 per cent adults (aged 18 and over). The notion of the videogame audience as comprised largely, if not exclusively, of pre-pubescent males has been remarkably pervasive in both popular and academic discourses. However, just a couple of years into its life, Sony's own demographic research suggested an average age of 20–21 for PlayStation users. More recent figures illustrate the continuation of this trend. The IDSA (Interactive Digital Software Association) offers a snapshot of the current US market. According to a survey conducted by Peter Hart Research for the IDSA and the *IDSA State of the Industry Report 2000–2001*, 60 per cent of all Americans (about 145 million people) play console and computer games on a regular basis. Moreover, the majority of console game players are 18 years and older. In fact, the IDSA claim that the average age of videogame players is 28 (see also Berens and Howard 2001: VII). The contemporary demographic suggests that the audience is comprised both of 'new entrants' discovering videogames anew and players growing up with the industry.

GENERATION PSX, MAINSTREAM AND HARDCORE: TARGETING THE AUDIENCE

Throughout the mid-1990s Sony, and to limited commercial success, Sega, attempted to market their consoles and games within the arena of popular club culture. This was part of an attempt to reposition video-games and appeal beyond the market of pre-existing players. The aim was to hail a then-untapped mainstream audience who had perhaps never played videogames before and may even have been resistant to them given the popular misconceptions of childishness and triviality. In order to achieve their commercial goals, it was necessary for Sony to refashion the videogame audience. Newman (2002b, 2001) has termed this mass-market audience 'Generation PSX' (PSX was the PlayStation's production codename and remains in popular parlance, although it is not used by Sony).

Apparently eschewing more traditional forms of marketing, and freed from the baggage of previous generations of consoles, Sony adopted an 'underground' strategy. At the Glastonbury Festival, they (in)famously distributed PlayStation publicity materials in a form that made them more than coincidentally suitable in the manufacture of hand-rolled cigarettes. Though the contents of the cigarettes were left to the discre-tion of the individual concert-goer and there has been no suggestion that Sony did or does endorse drug use, the association of PlayStation with the dominant youth culture of music, drugs, clubs and dance was surely not undesirable. For Geoff Glendenning, one of Sony's underground marketing gurus, the association with club culture and the avoidance of mainstream advertising was crucial to the PlayStation's initial success:

> I knew I had to get the underground magazines in, the people who are real individuals, get them on our side and create massive hype, and I needed to do that six months before launch. It had to be almost as if PlayStation was something they had personally discovered.
>
> ('nuGame culture', Edge 1996: 58–59; see also
> 'Hip or hype?', Edge 1996: 56–64)

In contrast, Nintendo has concentrated its marketing efforts on emphasizing the quality of its games. Certainly, the GameBoy Advance and GameCube have been positioned as pure gaming devices with none of the multimedia terminal aspirations of PlayStation 2 or Xbox. It has been noted that Nintendo takes great pride in its family image. For example, Nintendo's corporate website includes a 'parents' section with advice on how to select games for family play (see *Nintendo Information for Parents* and Sheff 1993). Many of Nintendo's games present apparently 'child-like' worlds and/or graphics. *Super Mario World 2: Yoshi's Island*, for example, is even presented in the style of a child's crayon drawings. While games such as *Goldeneye*, *Turok: Dinosaur Hunter*, *Perfect Dark* and *Resident Evil* are far from child-like in theme or aesthetic, Nintendo has managed to maintain the family-friendly image that Sony and Microsoft seem less keen to nurture.

Despite utilizing non-mainstream advertising and marketing techniques, the PlayStation's legacy has been the (re)creation of the mainstream or 'casual' gamer. The casual gamer can be seen as a direct consequence of the widening of the audience for videogames. However, while terms like hardcore and casual gamer have become part of industry parlance, as Rouse notes, quite who they are or what they want is rather difficult to ascertain. Certainly, usage is one means of demarcating the groups. Hardcore gamers are seen to be more committed to gaming as an activity and can be assumed to play more and more frequently, and as we shall see below, researchers have attempted to differentiate game preference and effects for high and low usage players. For many 'hardcore' gamers, the acid test is whether the interest in videogaming predates PlayStation. Interestingly, though perhaps unsurprisingly, the emergence of the casual gamer has affected games themselves. As Rouse (2001: 311–312) notes, a game like *Myth*, designed by and for hardcore gamers, is precisely the sort of title that many publishers might demand was simplified so that non-hardcore gamers were not intimidated by its 'complex controls or sadistic level of difficulty'. So, it is possible to suggest that the emergence of games as a mass-market, perhaps even mainstream, form has had an effect on the nature of videogame content and interfaces. Player-oriented magazines and websites are testimony to this, with self-confessed hardcore gamers deriding titles for pandering to the mainstream. Typical complaints found in the letters pages of videogames magazines or on online discussion boards (see *Edge Online*, for example) include 'oversimplification of controls', 'a general ramping down of overall difficulty levels', and 'ever-shorter games that require less commitment to play, complete or master'.

It should come as little surprise to find that, while the PlayStation family continues to enjoy enormous mass-market success, there is a feeling among some hardcore gamers that this success has impoverished the videogames they hold dear. Nintendo, with its SNES, GameBoy, GameBoy Advance and now GameCube, continues to woo the hardcore gamer with its 'pure gameplay' focus. Moreover, rather than attempt to gain credibility through associations with popular bands or other pop culture movements, Nintendo foregrounds its master game designer, or 'auteur', Shigeru Miyamoto, creator of *Donkey Kong*, *Super Mario Bros*, *The Legend of Zelda* and *Pikmin*, among others.

Scrutiny of web-based bulletin boards, fansites and community forums (see the discussion groups on www.gamefaqs.com, for example), throughout the 1990s, and taking advantage of the staggered release of videogames, self-appointed hardcore gamers delighted in the import gaming scene. Having games months before their official release in your territory, particularly if they were in the original Japanese, marked you out as a hardcore gamer immediately. It was helpful that territories using the PAL television standard (Europe and Australia, for example) were often treated to technically substandard conversions.

Technical differences between TV standards in Europe (PAL) and the US and Japan (NTSC), particularly in relation to the number of scan lines and different screen refresh rates, meant that considerable additional programming or optimization work was required in order to ensure the conversion ran at the same speed as the original and that graphics appeared the same. Effectively, the European videogame console had to work harder to render an image onscreen. Consequently, the many unoptimized conversions released in Europe ran approximately 17 per cent slower than their US or Japanese originals and suffered from 'letter-boxing'. Rather than reprogram the image rendering routines to take advantage of the additional scanlines on European television sets, graphics were framed top and bottom by black borders. Perhaps more problematically, because the horizontal scanlines on European PAL television sets are physically closer together than US or Japanese NTSC sets, graphics appeared 'squashed'. In Europe, Super Mario appeared even more squat than in Japan or the US.

In this way, even the same game could be seen to exist in hardcore or casual gamer form. Certainly, postings to fansites are highly critical of poorly optimized conversion of games like *Tekken 3* on PlayStation – sluggish response, slow-moving characters, compressed, letterboxed graphics and, worst of all, having to relearn how to time the complex series of moves mastered in the arcades. This apparent discernment is revealing and illustrates the ways in which the community of players evokes and reinforces categories such as 'hardcore' and 'casual' in order to justify pleasure and status within videogames culture and recalls the delineation of 'Trekkies' and 'Trekkers' (see Jenkins 1992 and Chapter 9 on fandom and videogame cultures). In a similar way, discussing pulp science fiction author E.E. 'Doc' Smith, Huntington illustrates the ways in which the form resists the efforts of the 'casual' reader, thereby elevating not only the status of the genre but also of the hardcore audience and their interpretive strategies and abilities. 'The casual reader does not understand science fiction, does not have sufficient animation or depth or breadth of vision to grasp it' (1989: 48).

While the late 1990s saw hardcore gamers attempting to grab the future of gaming by getting their hands on the latest titles, more recently, hardcore gamers have begun to look to the past for gaming experiences unaffected by consideration of the mainstream player. The emergence of retrogaming can be seen as an attempt to reclaim videogaming from the mainstream and can be understood as a form of hardcore fan resistance. Retailers sell consoles and games from the 1980s alongside the latest Xbox, GameCube and PlayStation 2 releases, and innumerable websites and fanzines champion the superiority – and particularly the difficulty – of classic videogames. The second-hand market and emergence of a retrogaming scene that highlights the history and heritage of videogames reinforces the cultural and social status of games as a medium and certainly represents an attempt on the part of marketers and retailers to appeal directly to hardcore gamers as a market segment.

BOYS ONLY? AUDIENCE DEMOGRAPHICS

As Provenzo (1991) has noted, adolescent males were a core market for videogame publishers during the late 1980s. While we have seen that the average age of players has steadily increased during the 1990s, particularly in the second half of the decade since the release of PlayStation, the popular perception of videogaming is that it remains a male-only pursuit. According to Mediascope (1999) 'Since most software is designed with boys in mind, the world of computing seems to be more consistent with male adolescent culture than with female values and goals.' Jenkins (2001: 4) concurs, 'Suppose we take at face value the claim that game designers aren't designing for boys – they are simply designing games they would like to play. The existing employee pool for the games industry is overwhelmingly male, so the games designed appeal overwhelmingly to men.' One consequence of this industry bias is the inequality of representation in videogames. Children Now's *Fair Play* report highlights the situation in terms of gender (im)balance:

- Female characters were severely underrepresented in video games, accounting for only 16% of all characters.
- Male characters were most likely to be portrayed as competitors (47%) while female characters were most likely to be portrayed as props or bystanders (50%).
- Male and female character roles and behaviors were frequently stereotyped, with males more likely to engage in physical aggression and females more likely to scream, wear revealing clothing and be nurturing.

('Children now' *Fair Play* 2001: 1)

Certainly, it is true to note that, where female characters exist at all, they are frequently relegated to the periphery or background and are not nearly as visible as their male, or even non-human counterparts. *Ridge Racer Revolution*'s 'Reiko Nagase' is a case in point (see *Official Namco Website* for more on Reiko). Serving no purpose other than to appear in box art and wave the chequered flag at the start of the race, Reiko is the epitome of outmoded and unwelcome stereotyping. Not only does the character serve no function in the game, and cannot be controlled by the player, she is the *only* character in the game. The cars that the player drives are just that – cars – with no drivers depicted. The *Super Mario* series' Princess character shows minimal development. In her earliest outings in *Donkey Kong* and *Super Mario Bros*, Princess was the archetypal damsel in distress. Constantly captured and requiring Mario to rescue her, she was the perpetual victim. Later games such as *Super Mario Kart* and *Super Mario Bros 2*, and recent titles such as *Super Smash Bros*, *Smash Bros Melée*, *Mario Tennis* and *Mario Golf*, however, have seen Princess emerge as a playable character with her own unique sets of techniques, capabilities, strengths and weaknesses. However, aesthetically, she has been unable to shake free of her long blonde hair and pink full-length dress (see Chapter 9 on fandom and the repositioning of 'Princess' through fan art and fiction, for example).

For many commentators, female players find videogames unappealing because of their apparent endorsement of gender stereotypes and their promotion of antisocial behaviour, such as violence, as a strategy for success. Additionally, as we have seen, there is a paucity of the kind of characters in which female players might invest:

Examine the game box of one of the top-selling games. You will likely see a male with a weapon, and possibly a well-endowed damsel in distress. If you happen to choose 'Tomb Raider,' you will see an underdressed buxom female with a weapon. Both pictures are likely to be attractive to males, and

[are] basically consistent with schemas for male behavior (being aggressive or scrutinizing good-looking females). Neither picture is likely to be appealing to females because they invoke the gender stereotype that women are brainless and helpless or that power for women is dependent on sexual appeal . . . If you choose to play one of the top-selling games, you will likely be required to use ruthless competition and violence to succeed. For males, these activities have considerable social approval. For females they do not.

(Funk 2001: 1)

Interestingly, despite indefensibly stereotypical representations, recent US player surveys suggest that women may be more involved with videogaming than is popularly reported. The IDSA has claimed that females exert a considerable influence in hardware purchasing decisions: 'About one-quarter of those with the most influence over console purchases are female; while 41% of all influencers are age 18–35, 27% are over 36, and 32% are under 18' (*IDSA State of the Industry Report 2000–2001*). Perhaps more controversially, in their 'Top ten industry facts' (2002), the IDSA has also claimed that 43 per cent of US game players are women. Some critics, including Jonas Smith (2001a), have questioned the validity of the findings, and extrapolations based on them. The undisclosed methodology of the IDSA study leads Smith to claim that the data could support two wildly different conclusions: first, 'women play as much as men', second that 'A more or less equal number of women and men have played computer games at some time. Men, however, spend X times more time and money playing.' For Smith, the data have been stretched to breaking point and, while questionable in themselves, are clearly being used to serve the agendas of different groups. Henry Jenkins reminds us that the sight of a female player taking on, and beating, allcomers at a trade show remains remarkable suggesting that videogaming remains a commercial and cultural space dominated by male designers and consumers (2001: 1). However, for many, the IDSA's figures have been a great fillip and, according to Eisenberg (1998: 3) 'For the millions of women who grew up playing video games like *Space Invaders*, *PacMan*, *Frogger*, *Pong*, *Tetris* and *Centipede*, the "fact" that girls don't play games should come as quite a surprise' (see also Smith 2001b).

The industry is keen to capitalize on this market and two broad strategies can be identified. The first is to make games with wide appeal – 'crossover' games (or perhaps 'gender-neutral' games), while the second is to directly target the female market with 'games for girls'. According to Eisenberg, with games like *Sonic the Hedgehog*, Sega has deliberately set about designing videogames to appeal to the broadest possible audience. Sony, too, see crossover appeal as essential and Eisenberg points to the success of titles such as *Final Fantasy VII* and *Parappa the Rapper* in Japan where the consumer base is said to be 40 per cent female (Eisenberg 1998: 3). In contrast to the inclusive aims of crossover games, 'games for girls' seek to appeal explicitly and solely to a female audience. These so-called 'pink games' are often aimed at young girls and include titles inspired by existing franchises like Barbie or Disney's Little Mermaid and Beauty and the Beast:

> Armed with research they say proves girls don't like the 'overly competitive' games that 'boys like', and fueled by the success of Mattel Media's CD-ROM game *Barbie Fashion Designer* last fall, new companies like Purple Moon, the maker of *Rockett* [Rockett's New School], have come onto the market to attract the dollars of young girls.
>
> (Eisenberg 1998: 1)

Brenda Laurel, founder of Purple Moon, points to her four-year research project profiling the play preferences of girls that included interviews with girls and boys in 'friendship pairs', arcade managers, schoolteachers and camp counsellors, to suggest that girls need 'friendship adventures' while boys need 'action games'. However, Eisenberg (1998: 2) has questioned whether, ' "Friendship Adventures" based on popularity and clothing are really what girls deserve?' Critics of girl games suggest that by focusing on popularity and fashion, the content of these games can do little to redress any gender imbalance in the gaming world. For Adams, the focus of girl games is problematic and limiting and does little more than reinforce unhelpful and outmoded stereotypes:

> Boys get to drive Formula race cars, fly F-15s, build cities, battle dragons, conquer the galaxy, save the universe. Girls get to . . . become queen of the prom? Is that really the best we can do for them?
>
> (Adams 1998: 2, original ellipsis)

In fact, it is argued that the very notion of games for girls is problematic in itself. Adams suggests that the problem goes further than the videogames industry:

> Most companies organize themselves along functional lines. The top-level divisions in a company are usually things like marketing, sales, R&D, manufacturing, distribution, and so on. Not so with toy companies. They divide themselves, right form the start, into 'boy' divisions and 'girl' divisions.
>
> (Adams 1998: 3)

Adams argues that this makes no financial sense as placing 'for girls' or 'for boys' on the box immediately cuts your potential market in half. However, there is an even more serious implication. Because the majority of videogames are not explicitly marketed for one sex or another, marketing certain videogames for girls ghettoizes them:

> It's not empowering, it's limiting. It reinforces the notion that femaleness is a special case, an exception to the norm. If the number of games for girls is a tiny fraction of the total, it tells the girls that they're second class cyberci-tizens, who have to make do with what they're given . . . Why make 'games for girls'? Why not just make *good* games for everybody?
>
> (Adams 1998: 3–4)

Muller (1998) has noted that for many girls, pink games are simply too 'girly-girl' and certainly the existence of all-female *Quake* clans like 'Psychotic Men Slayers' (PMS for short) goes some way to confirming this (see Brown 1997, for example). Through their rejection of pink and crossover games, and by virtue of their direct engagement with 'boys' games', these 'Quake Grrls' have been viewed by many as examples of women players' resistance to industry's prescribed notions of femininity and appropriateness and an important step toward claiming video-games as a genuine space for girls. However, as Cassell and Jenkins (1998: 26) note, 'The "Quake Grrls" are, on the whole, older than the girls being targeted by the girl games movement, more self-confident, more comfortable with technology, and more mature in their tastes and interests.' Moreover, 'Quake Grrls' do not challenge the underlying norms of demarcation. These are still 'boys games' and it is possible that 'Game Grrrls can always be read as a harmless aberration' (Cassell and Jenkins 1998: 27).

Both the industry-derived 'games for girls' and the player-led 'Quake Grrls' movements can be seen as counterproductive as they marginalize or ghettoize the female player in what remains a male realm. The situa-tion faced by the contemporary videogames industry is clearly problematic. If the prevalence of stereotypes contributes to the non-participation of women, then tackling the representational imbalance is essential. The challenge is to avoid the overreaction of the pink games movement and find ways in which videogames can both shake the representational baggage of their past, and encourage players in a less blunt manner. Inevitably this will have to involve factors beyond the interaction with the game despite the potential pitfalls of gender ghet-toization. Videogames do not exist within a vacuum. Rather, they reside, are produced, and are encountered within a web of intertextuality in which explicit and implicit references to other media forms proliferate in videogames, and in which videogames are referred to aesthetically and stylistically within other media. As such, advertising and marketing materials, not to mention the various and extensive tie-ins and spin-offs such as movies and cartoons, must be considered alongside the content of the games. As Bennett and Woollacott (1987) have noted, materials such as these play an important role in actively cueing readings in particular ways and encourage us to consider the action of more than merely 'the text' (see Chapter 8).

JUST FIVE MORE MINUTES . . .
MEASURING AUDIENCE BEHAVIOUR

To understand videogames, it is not just important to know the issues of gender impinging on them. Other factors in respect of *quality* of gameplay need to be taken into account. How many hours a week do players spend playing videogames? Every day? Do players even know how many hours they spend playing? The way in which videogames are played and the way in which the content is delivered clearly marks the form out as significantly and substantively different to watching film or television or listening to the radio. Television, radio and film content is generally distributed and portioned out into 'programmes', 'episodes', or 'features'. As we shall explore in the following chapter, videogames, are usually segmented into levels. Typically, levels do not take a specific period of time to complete, and thus the total duration of the level may be as dependent on the player's ability as on the content of the game per se. However, not all games present a free-roaming experience and most include some form of timer providing an additional impetus to complete speedily. Furthermore, even where no explicit timer or counter is incorporated, some games have relentless autoscrolling that effectively pushes the player through the level forcing and controlling the pace of the experience. That is, in a horizontally scrolling game like *Super Mario World*, for example, where normally a player can traverse a level at their own pace and effectively control the rate of scrolling, some levels may automatically scroll thereby forcing the player through the level at a pace dictated by the game. Similarly, driving games often require the player to complete a finite number of laps. Yet, even in these situations, it is not always possible to predict the amount of time it might take to play through or complete a particular level.

Remember also that it may take the player many attempts to complete the level. Even what we mean by complete has a bearing on the amount of time taken. For example, does completion mean to truly finish, or merely register a performance to the satisfaction of the player, or beat a previously set high score or time? Again, it is important to note the complexity and potential fluidity of rule systems and objectives operating with any given videogame play session and players are free to engage in open-ended paidea or impose their own ludus rules upon the frame of the game. Moreover, trying to beat a three-minute time-limited level can still result in many unplanned hours of play into the small hours of the night.

Though research studies show quite varied results, the consensus of studies investigating the incidence of play is that videogames are a high-frequency activity. Funk's (1993: 89) US study found that two-thirds of girls and 90 per cent of boys played for at least 1–2 hours per week with some reporting in excess of 15 hours. In the UK, Granada TV's 1993 'World In Action' programme surveyed 150 13–15-year-olds and found that nearly half were playing up to 25 hours per week. Perhaps more significantly, Philips *et al.* (1995: 687–688) note that individual playing sessions may be long (and frequently longer than initially intended) with 75 per cent at over half an hour and 14 per cent over 2 hours at a time. Drotner (2001) shows that videogame play is not only a frequent activity with long average play sessions (44 minutes daily average for UK 9–16-year-olds) but that it is by far the most prevalent computer-based activity, as non-game computer use clocks in at just 30 minutes per day on average with Internet use just 10 minutes per day.

Quantifying patterns of play is somewhat difficult as research necessarily lags behind the current marketplace; Drotner's 2001 study uses data gathered in 1998, for example. Moreover, despite the contemporary videogame industry's demographic, research tends to be focused exclusively on the playing patterns of young children. This is perhaps understandable given the motivation for much of this work. Many scholarly studies of patterns of play are stimulated by a desire to ascertain what activities the videogame is replacing. As such, there is often a barely concealed distrust and dislike of videogames underpinning the work (note the reviews presented in Emes 1997; Kline 1999; Provenzo 1991).

Nevertheless, there are other important factors in the quantifying process. Kline (1999: 20) has noted that compared with light or occasional players, 'heavy players were more likely to put off doing homework and chores (37 percent) and family activities (18 percent) than leisure activities (13 percent) or spending time with their friends (10 percent)'. Quite whether computer games occupy a unique position in being viewed as more desirable than homework or household chores is probably debatable, yet the fact that players, like the participant in Kline's earlier study (1997) who claim that they 'could play video games for hours and not notice' (cited in Kline 1999: 20) clearly indicate the potential for unwitting substitution of gaming for other activity. Moreover, Egli and Meyers (1984, cited in Dill and Dill 1998: 409) have noted that 13 per cent of the adolescents they surveyed 'sacrificed other attractive activities so that money and time could be devoted to video-game play'. That players can be seen to be skipping homework in order to spend time playing games has led some researchers to postulate a negative correlation between game play and educational performance. However, here also, results are inconclusive:

'In a study of eighth graders, frequent video game players were the heaviest television viewers and performed the most poorly at school' (Lin & Lepper). However, '. . . a study of college psychology students, no differences existed between frequent and infrequent video game players on measures of class attendance, locus of control, or grade point average' (McCutcheon, L.E., & Campbell, J.D.).

(Mediascope 1999)

Additional concerns arising from the high level and frequency of play centres on the notion that computer game play is essentially sedentary. Clearly this debate can be located within a broader discussion of the increasingly sedentary lifestyle of young people engaging with various media forms such as television and VCR, and latterly DVD and the Internet. The videogame, in this sense, can be seen to contribute to the creation of the 'couch potato', and it is from this point that many studies of computer gaming begin. However, as Segal and Dietz (1991: 1034) have noted, the direct causal link between computer game play (and even television viewing) and the incidence of obesity remains contested. In fact, Segal and Dietz's study makes what may be a surprising distinction between television viewing and computer game play that is based around the marked differences in physiological effects. While they are at great pains to ensure that it is not considered a substitute for intensive physical exercise as the cardiorespitory stress involved is not sufficient to increase fitness, Segal and Dietz note that:

The primary finding of this study was that video games are not a passive activity and the energy cost of the game approximates mild-intensity exercise . . . the effects of playing video games on energy expenditure suggest that it is not a passive activity.

(Segal and Dietz 1991: 1034–1035)

The study is interesting in illustrating a high level of active engagement with the videogame system (approximate 25 per cent increase in heart rate exhibited by all subjects during engagement, for example) and the authors mark the difference between this and the passive engagement of 'viewing' and 'observing' associated with television (see also Graybill *et al.* 1987, 1985). Although the study can hardly be considered conclusive, its findings are useful in highlighting the more general and recurring theme of this book specifically, and videogame studies in general, as to the potential dangers of the simple transference of conceptual models, or working on the basis of presuppositions, derived from studies of other superficially similar media forms. Further problems arise from the specificity of studies of patterns of play. Not only are the majority of findings confined to particular territories, with much work emanating from the US, but sample groups are often quite locally-delimited. Extrapolation from extant findings is potentially dangerous, and it is clear that more research is required so as to build up a better, and more globally representative, picture of videogame play.

GAME PANICS: 'EFFECTS' RESEARCH AND THE INSCRIBED AUDIENCE

There is a reasonably long history of academic research into the effects of the media on their users. For decades, scholars have focused their attentions on a range of different media from penny dreadful comics, through nineteenth-century amusement houses, popular theatre, to cinema, television, comics (again), video (particularly so-called 'video-nasties') and the Internet (see Cumberbatch 1998: 262). Most major popular cultural forms have been subjected to the analyses of critics seemingly determined to highlight their negative properties and the harm they inevitably will cause both the individual and society if unchecked through some form of censorship or control. The Internet is perhaps the most immediately obvious example of concerns over the damaging, antisocial effects leading to technical, legal and social mechanisms designed to limit the damage (see Hunter 2000). As Bourdieu (1990) has observed, the denigration of the popular may be understood in terms of its impenetrability. Consequently, popular forms are frequently presented as uncouth, dangerous and harmful by those lacking the knowledge and strategies to make sense of them. Yet, this is not just a perspective on the nature of such forms as 'texts' or artefacts. Importantly, it also embodies a strident political perspective on the psychology – primarily the susceptibility to suggestion – and intelligence of audiences. As we shall see, both in this chapter and others in this book, there is a considerable effort to position videogames as harmful. In considering attempts to link videogame play with horrific events such as shootings and putative societal decay, we can note the politicization of the process of denigrating popular culture.

As Cumberbatch (1998) has noted, most of the effects research derives from the laboratories of psychology departments. By contrast, media and cultural studies scholars have tended to reject the underlying causal mechanism (see Gauntlett 2000, 1998, 1995, for example) particularly in terms of its limited scope. Assuming that the effects on readers may be deduced from scrutiny of media texts themselves, such studies have been criticized as falling foul of what Thompson (1990) terms 'the fallacy of internalism' and fail signally in considering the contexts and practices that characterize media use and the interpretive practices, such as fandom, through which meaning is generated (see also the body of research on audience 'activity' including Ang 1996, 1985; Hermes 2002, 1995; Morley 1986, 1980; and see also Chapter 8). Moreover it is generally true to suggest that effects studies have tended to focus not merely on popular cultural forms, but specifically on children's media – or at least the effect of media on children, thereby highlighting two further weaknesses.

Some studies simply maintain the conceptualization of the videogame audience derived from 1970s' and 1980s' demographics, while others apparently imply the audience from the text. As such, the cartoon aesthetic of *Super Mario Bros*, for example, may be read as an appeal to young players. Furthermore, consideration of the structure of the videogame text, and in particular the ways in which games encourage replay and re-engagement, is frequently read as signalling their addictiveness. The consequent positioning of the player as 'addict' provides a convenient, if poorly justified, means of explaining both the high levels of play and the propensity of heavy players to skip other activities or duties in favour of game play. For example, Klein (1984: 396) noted that, in order to get their 'video game fix' and feed their addiction, many of the children he counselled had cut classes, spent their lunch money or even begged or stolen money, while Braun and Giroux (1989: 101) have seen computer games as 'the perfect paradigm for induction of "addictive" behavior'. It is a small but significant step to move from the discussion of addictive or, perhaps, compulsive behaviour to the treatment of computer games as a drug. However, the position is not uncommon and serves the purposes of those seemingly determined to vilify games and gamers.

Zimbardo (cited in Klein 1984: 397) was in no doubt as to the gravity of the situation, 'Video game fanatics essentially are like cocaine addicts who get an instant rush from an electronic fix.' More recently, Prof. Lt. Col. Dave Grossman has employed similar language to describe what he considers one of the blights of contemporary society. While, as we will learn, Grossman's principal interest concerns violent content, the language and terminology he uses encourages us to consider the video-game as a malevolent force. Players are not players but rather are described as 'users' – users of a drug, presumably 'pushed' rather than sold (Grossman 2001, see also 'Video nasties', Channel 4 Television 2000). Discussions of computer game addiction such as those illustrated above, really present little more than anecdote and must be considered to employ the terminology of 'addiction' in its loosest possible sense (note also the discussion of intrinsic reinforcement schedules in Loftus and Loftus 1983: 10–33). Similarly, allusions to the narcotic effects of computer games must surely be treated as metaphor, and quite probably reveal a highly politicized stance, as is doubtless the case with Grossman, for example.

Though anecdotal reports precede the formal recognition and discussion of the condition in 1990, 'Nintendinitis' refers to the short-term injury affecting the player's right thumb as a consequence of the repeated button pressing (Dorman 1997). More recently, and perhaps reflecting the shift in dominance in the global industry, the same condition has become known as 'PlayStation thumb'. In addition to the more usual comparison of high scores, progression through levels, or lap times, fansites play host to (admittedly light hearted) discussion and comparisons of calloused hands as indicators of prowess and dedication! Adding to this more general and short-term injury, more serious calls have come from doctors in the UK for additional health warnings to be added specifically concerning game controllers (cf. 'Vibrating games health warning', BBC News Online 2002). Though predated by coin-op implementations (for example, Sega's *Virtua Racing* and *Daytona USA*) and deriving from military applications (such as rumble seats in flight/tank simulators), Nintendo's 'Rumble-Pak' (N64), Sony's 'Dual Shock' and 'Dual Shock 2' (PSone and PS2 respectively) and Sega's 'Vibration Unit' (Dreamcast), in addition to various third-party devices for consoles and general purpose PCs alike, each brought positive feedback to the home gamer. It is important to understand that such devices typically offer some form of tactile or haptic feedback via a vibration unit built into an otherwise standard controller, such as a joypad. The effect of the various levels of momentary vibration is to reinforce the action on-screen, so driving over a rumble strip at the edge of a racing track generates a different *feeling* response to crashing into a barrier or another vehicle.

Though they have been treated to nothing other than praise among the gaming community since their introduction in the mid-1990s, calls for additional health warnings came after a teenager developed a condition known as hand-arm vibration syndrome. Over a two-year period, the 15-year-old boy had suffered pins and needles, and his hands had became white and swollen when exposed to the cold, and red and painful upon warming. The symptoms are typical of those more usually experienced by those working in industrial situations using, for example, chainsaws or pneumatic drills. However, in this case, the symptoms were attributable to the use of a vibrating game controller. In response, Sony claimed that, since the introduction of the PlayStation in the mid-1990s, it had never before received any feedback or complaints regarding hand-arm vibration syndrome as a result of using feedback devices. Moreover, Nick Sharples of Sony Computer Entertainment Europe observed that, 'Given the vast number of people who have had these games and the lack of reports of any problems, perhaps it's not necessary to have the warnings' ('Vibrating games health warning', BBC News Online 2002). While this might initially appear a somewhat dismissive or complacent attitude, it must be noted that the boy reportedly played for seven hours each day. In addition to the potential health concern, perhaps this example serves most powerfully to highlight again the high levels and rates of play.

As Ivory (2001: 5) notes, 'studies involving the physiological health-based pros and cons of video game play use by children fail to present a clear picture of video games as a positive or negative influence'. So, while concerns mount over the amount of play and the possible detrimental health issues, a number of physiological studies present computer games as beneficial. Research has particularly concentrated on the correlation between computer game play and enhanced motor skills. Orosy-Fildes and Allan (1989) have demonstrated decreases in reaction times in both male and female players as a consequence of game play, while Kuhlman and Beitel (1991) have correlated game play experience with players' (of both genders) ability to perform predictive motor tasks (using a Bassin Timer). In addition to the more general hand-eye co-ordination improvements that are noted in these studies, the findings indicate that computer games may provide an effective means of challenging the gender variances that are usually noted in motor skills. Often, it is difficult to establish whether such 'effects' of videogames are 'physical' or 'emotional' in their bearing. Interestingly, Sony have been very keen to announce the arrival of 'emotional gaming' with PlayStation 2. Indeed, they have dubbed the graphics chip the 'Emotion Engine'. It is curious to note that while the company has been understandably keen to decry effects research not least through its PSone UK television advertising, the idea of videogames 'affecting' players is central to its more recent strategies (see www.playstation-europe.com).

It is regrettable to note that a number of tragic events, perhaps most notably the Columbine High School shootings, have heightened concerns over videogames and their effects. On 20 April 1999 in Littleton, Colorado, Eric Harris and Dylan Klebold killed 13 people, wounded a further 23 before turning their guns on themselves. In the media discussion that followed, it emerged that Harris and Klebold, like Michael Carneal, who in 1997 had killed three of his classmates in a school shooting in Paducah, Kentucky, were videogame enthusiasts. The link seemed clear. For some time, effects experts had been discussing the danger of violent computer games and here was proof. Michael Breen, attorney in the case taken out against Midway Games by the families of three of the students killed by Carneal, claimed that it was computer games that had helped to train the young killer:

> Michael Carneal clipped off nine shots in a 10-second period. Eight of these shots were hits. Three were head and neck shots and were kills. That is way beyond the military standard for expert marksmanship. This was a kid who had never fired a pistol in his life, but because of his obsession with computer games he had turned himself into an expert marksman.
>
> (Hanson 1999: 15)

Responses to violent computer games, like so-called video-nasties before them (Baker 1984), sometimes tend towards the hyperbolic and sensational. The apparent link between computer games and the school shootings highlighted in the previous paragraph has prompted Dave Grossman to ask 'Are we training our children to kill? . . . Every time a child plays an interactive video game, he is learning the exact same conditioned reflex skills as a soldier or police officer in training' (2001: 3/6). Violent computer games, according to Grossman, can be seen to contribute to the creation of a society which is infected by:

> a phenomenon that functions much like AIDS, which I call AVIDS – Acquired Violence Immune Deficiency Syndrome . . . once you are at close range with another human being, and it's time for you to pull that trigger, Acquired Violence Immune Deficiency Syndrome can destroy your midbrain resistance.
>
> (2001: 6)

ASSESSING THE RESEARCH

Presenting a summary of the extant research into the effects of violent videogames is problematic in itself as the findings of the various studies, as Kline (1999) and Griffiths (1999, 1997a, 1997b, 1993) have noted, are inconclusive and often contradictory. Moreover, as Griffiths and others have observed, methodological flaws blight many of the studies. The lack of consistency in the selection of participants reveals something of an imbalance in favour of very young players. This makes assessments of the developmental nature of any identifiable effects impossible, and further highlights the need for longitudinal studies. Furthermore, many studies operate comparative tests using various classifications of games as 'violent', 'non-violent' or 'medium violent' (Schutte *et al.* 1988; Cooper and Mackie 1986). The basis for such classifications is frequently spurious, undermining the validity of subsequent findings. Scott (1995) notes that supposedly non-aggressive applications in the Graybill *et al.* study (1987, also 1985) did not represent particularly good examples and reports Cooper and Mackie's observations that the (female) participants in their study 'saw little difference between Pac Man and their aggressive game, Missile Command' (Scott 1995: 123). Though distinctions between 'aggressive' and 'non-aggressive', 'violent' and 'non-violent' videogames are used freely in the various studies, there is no consistency in the definitions. In some studies, violence and aggression are measured in terms of representation, hence violent *Doom* and non-violent *Super Mario*, while in others, the required techniques and strategies are the yardstick, rendering both *Doom* and *Super Mario* aggressive and violent.

It is possible that definitional issues may help explain some of the differences in the perceived prevalence of 'violent games' (see Funk 2001). Although games such as those in the *Super Mario* series require the player to destroy the creatures that bar their way by, for example, jumping on top of them, few if any, players would consider these titles to be 'violent', and they are not cited by even the most vocal critics of games violence, such as Grossman, as requiring censure. Chory-Assad and Mastro (2000: 3) have similarly observed, 'Using today's standards, "Pac-Man", and other early videogames like "Space Invaders", "Defender", and "Asteroids" appear relatively non-threatening; however, in the early 1980s these games were characterized as violent.'

Even if we assume, momentarily, that it is reasonable to consider the action in *Pac-Man* or *Super Mario Bros* as 'violent', then surely we cannot simply equate it with the type of content and action we see in *Mortal Kombat*, *Goldeneye* or *Soldier of Fortune*. The inconsistent definition of 'violence' and 'aggression' renders it difficult to compare the findings of studies. Moreover, the issue of narrowness is raised as studies using just two games take little account of player preference and risk erroneous extrapolation from potentially unrepresentative games. While they acknowledge the existence of non-violent videogames, it is generally true that effects studies have tended to overemphasize the prevalence of violent content and action. It is notable that driving and puzzle games include no representational violence and are better understood as promoting competitive rather than aggressive behaviour (Anderson and Morrow 1995), yet enjoy enormous commercial success (note also the two top selling PC titles, *Myst* and *The Sims*).

Emes (1997) has also pointed to the potential unreliability of some measures of participant aggression citing the Calvert and Tan (1994) study that examined the responses to passive observation and active engagement with a 'virtual reality game' in terms of 'aggressive thoughts'. Though the study suggested a link between active participation and the prevalence of such thoughts, Emes (1997: 412) notes that, 'Having aggressive thoughts after playing video games does not necessarily translate into aggressive behaviour.' Furthermore, Sheppard (1997: 489) notes that many children appreciate the context of games and enjoy the excitement and participation without demonstrating aggressive behaviour. In this light, the comments of Zimbardo (1982) appear even more bizarre and unjustified, 'video games are so addictive to young people that they may be socially isolating and may actually encourage violence between people' (cited in Scott 1995: 121–122), while the US surgeon general expressed the (personal) view that, children 'are into the games body and soul – everything is zapping the enemy. Children get to the point where when they see another child being molested by a third child, they just sit back' (cited in Scott 1995: 122).

The use of 'fantasy' rather than 'behavioural' tests, as in Graybill *et al.*'s (1985) and (1987) studies further problematizes findings. As Freedman has noted, while they might be the only choice in ethical terms, the use of 'analogues of aggression', such as the administering of a loud noise to another subject, as in the Anderson and Dill (2000) study, is a potential area of concern, as such action can be deemed 'pretty remote from real aggression' (Freedman 2001: 8). Moreover, disentangling computer games from other aggression-inducing sources is problematic: 'it is difficult to isolate the role of playing video games from other contributing factors' (Emes 1997: 410). Scott's study concludes that 'individual variability [of participant] is more important than variability in affect induced by playing computer games . . . Some people may be able to spend a great deal of their free time playing arcade videos [*sic*] without any resulting aggression' and suggests that 'glib statements relating aggression to game playing, whether appearing in the mass media or scientific journals, seem totally unwarranted' (Scott 1995: 129–130 and 130).

The potential development of psychopathological disorder through extended use of videogames appears to have shaped the presuppositions of investigation into the field. However, as Funk (1992: 53) notes, 'researchers have failed to identify expected increases in withdrawal and social isolation in frequent game players'. Emes concurs, citing the Kestenbaum and Weinstein (1985) study which, although highlighting different 'uses' and reasons for playing between 'high-' and 'low-rate' players, 'concluded that heavy video game use did not result in global psychopathology or social introversion' (Emes 1997: 412). Sherry et al.'s (2001: 12) findings are more forceful still, in their discussion of the issues of 'isolation', 'introversion' and pathology: '[computer] gaming appears to be a type of diversion that involves other people in social interaction. This finding contradicts the idea of the solitary game player isolated from social contact. In fact, frequent game play appears to be highly social' (see also Chapter 9).

Graybill et al. (1985: 204–205) noted their observed postplay differences may have been attributable to relative game difficulty thereby stressing the significance of interactive experientiality over purely (visual) representation. This recalls Anderson and Morrow's (1995) assertion that the competitiveness of the situations may be largely responsible for any observable post-play aggressive feeling. Griffiths (1997a) takes a slightly different stance that questions the validity of the underlying effects mechanisms and, specifically, their unidirectionality. Citing Fling et al. (1992) and Griffiths and Hunt (1993) whose studies both highlighted highly significant correlations between frequency of computer game play and self-reported aggressiveness, Griffiths (1997a) notes that correlational results such as theirs could indicate that more aggressive children are drawn to video games rather than – or in addition to – their aggression being a result of this activity.

Covering the areas of 'physiologic responses', 'videogame-related seizures', 'videogames and aggression', 'psychopathology' and 'academic performance', Emes provides a useful summary of the extant research into the effects of violent videogames on their players, in concluding that:

- The number of well-conducted research studies in this field is small
- The reliability and validity of the procedures used to measure aggression is questionable
- Research into the long-term effects of playing videogames is lacking

(Emes 1997: 413, original formatting)

For Jessen (1998, 1996, 1995), the failure to encounter videogames on their own terms is a significant one and reveals a lack of understanding of what videogames are, how they function, how they are used by players, and the contexts in which they are used:

> computer games primarily acquire their meaning and content through their concrete use in concrete situations. In this sense they are more a kind of tool for social relations than a means of communicating the messages one normally looks for in the media . . . we cannot interpret a content outside the concrete practice which also provides the framework of understanding. For example, something that might on the face of it look extremely violent on the screen may in practice have quite a different function. The players might for example blast one another and everything else in a violent game like 'Doom II', while at the same time enjoying extremely peaceful, playful relations, as is in fact usually the case in war games.
>
> (Jessen 1998: 43–44)

As we can see, then, glib assertions of what videogames *are*, based on beliefs about the way that they are played, are problematic. The core of the problem seems to consist of the theorization of the players as a 'target market' and the idea (often employed by academics) that the players of videogames – their attitudes, values and experiences – can be 'read off' from a cursory glance at such videogame 'content' as (undefined) violence. This can tell us little about videogames and, indeed, frequently contributes to a myth that occludes further scrutiny of the medium in question. In the next chapter, then, we will proceed to a close analysis of the *structure* of videogames which facilitates play.

5

VIDEOGAME STRUCTURE
Levels, breaks and intermissions

NON-INTERACTIVITY IN THE INTERACTIVE VIDEOGAME

It is curious to note that scholars in the field rarely consider the structure of videogames and, as a consequence, it is poorly understood. This is unfortunate as the implications of videogame structure are considerable. An understanding of the ways in which videogames are assembled, presented and experienced is extremely revealing and adds considerably to the identification and analysis of the audiences for gaming and, importantly, greatly assists in defining what it is to 'play' a videogame and the nature of activity/interactivity. As such, the discussion that follows here foreshadows issues that are revisited in more detail in subsequent chapters. The aim of this chapter is to highlight some frequently overlooked or misunderstood principles of videogames so as to provide a foundation upon which to build our analysis. Key among these is a (re)evaluation of the significance of the cut-scene.

We have noted in Chapter 2 that for many commentators, the videogame is defined by its focus on player *activity* (Livingstone 2002; Rouse 2001) and that videogames that offer limited potential in this regard are frequently decried (Waters 2002; Juul 1999). We might, then, think it logical that any sequence that offers little or no potential for such player activity might be at best superfluous. Certainly, the 'non-interactive cut-scene' has generated considerable consternation among players and commentators alike with some critics, such as Rouse, suggesting that their deployment undermines the work of the designer as the player is wrenched away from participative involvement in order to watch a pre-rendered 'movie' sequence. Design assumptions such as these posit the cut-scene as injurious to the integrity of the videogame experience as a whole. Yet, the simple fact is that they persist and may be found in games lauded with considerable critical acclaim such as Nintendo's *The Legend of Zelda: Ocarina of Time*. Moreover, the cut-scene is not alone in presenting sustained 'non-interactivity'. Breaks and intermissions between various 'levels' have been a staple of videogame design since the 1970s. As such, while the contemporary cut-scene may present a spectacle of audio-visual panache unimaginable even a decade ago, it should be considered responsible for the segmentation of videogames into 'interactive' and 'non-interactive' sequences. At least part of the aim of this chapter is to interrogate the popularly held notion that cut-scenes or other 'non-interactive' sequences undermine the integrity of the player's experience. In doing so, the discussion presented here aims to show that such materials, and the experience of them, may be considered to be central to the videogame experience rather than merely peripheral or even counterproductive.

In Chapter 2, we noted the difficulty of defining the object of study. Assembled under the banner of 'videogames' is an almost impossibly diverse range of technologies, experiences, game types and aesthetics. However, the difficulty is compounded if we scratch the surface of these titles. Simply, videogames rarely present a singular experience throughout. Whether we consider the current generation of Xbox, GameCube, PlayStation 2 and GameBoy Advance titles, or a vintage coin-op machine from the 1970s, videogames are highly structured, or perhaps more accurately, segmented. At the most fundamental level, videogames are almost always portioned up into levels, stages, rounds, bouts, sub-levels, mini-games, 'Boss' stages and so on (see the discussion of Nintendo's Game & Watch *Parachute* on p. 79 for an example of an unusual partial exception). Perhaps even more importantly, these different stages frequently offer different types of setting, action, location and even representational styles. In fact, individual videogames are frequently praised by reviewers and players alike for the variety of their gameplay types. A given videogame may offer a car driving level, followed by a hand-to-hand combat level, for example, or may, as *Grand Theft Auto* attempts, integrate these types of action and experience within single levels allowing the player to dynamically switch between them. The videogames industry has very strong incentives for creating such a situation. Keen to promote the longevity and value for money of their products, the diversity of experiential potential is a key weapon in the marketing armoury. As such, the structure of, and diversity within, videogames problematizes attempts to delineate videogames from one another along the lines of the type of action they present. That these characteristics are not necessarily fixed or consistent within individual games undermines the neatness of classifications based around content such as those presented by Berens and Howard (2001).

While it is evident that such characteristics are potentially inadequate in delineating specific titles, it is important to note that generic conventions abound within videogames whether in terms of aesthetics, functionality or control methods, and players' expectations are clearly couched within their ongoing experience of other titles. Indeed, it is difficult to imagine that such anticipation and experience is not demanded by videogame designers. To know that an enemy flashing white when hit by the player's weapon is actually taking damage and not blocking or evading in some way is essential in facilitating the formulation and evaluation of an attack strategy in *Metroid Fusion*, for example, but is explained nowhere in any instruction manual or in-game tutorial. This important feedback mechanism that allows the player to explore and test tactics is precisely what facilitates the interrogation of the simulation underpinning the game (Friedman 1995) yet it is not the invention of that game. The knowledge and expectation that a specific enemy will not only have a weak spot rendering them susceptible to a particular attack pattern but also that successful and unsuccessful strikes will be visually differentiated is gained from experience of other videogames. It follows that, by implementing such a generic feature, *Metroid Fusion*, just like *Super Mario Sunshine*, *House of the Dead 2* and *Halo*, places itself explicitly within a context of other titles and encourages the player to draw upon their extant knowledge and experience in tackling the problems laid out in this simulation model. Moreover, its application crosses 'fighting', 'shooting' and 'platform' games, for example.

The diversity of activity and experience engendered by a single videogame can be noted by examination of Sega's *18 Wheeler*. Ostensibly, this appears to be a racing game, with the unique selling point that the players race trucks rather than the more typical racing cars, go-karts or anti-gravity spaceships. However, this would be to ignore important sections of the game. Upon successfully transporting their chosen payload to its destination and beating their rival trucker, the player is presented with a Bonus Round. Here, racing is far from the agenda. Instead, the player is required to park their truck in a designated bay after having first negotiating the narrow, twisting backstreets. It follows that, as the player progresses further through the game, these backstreets become ever more labyrinthine, and time limits ever more demanding. Even this simple distinction between stages alerts us to the potential variety of gameplay offered within a single videogame.

However, for all their different demands, both of these stages are still concerned with driving. If we consider the so-called 'platform game', we find even more pronounced dissimilarities between sections or levels. The very name 'platform game' suggests a particular type of action – jumping across platforms to get from the start of a level to the end – yet, even defining the gameplay with this level of simplicity reveals 'platform game' to be a misnomer. How can it be that in a platform game the player frequently finds themselves racing in a minecart, or flying through the skies either in an aircraft or because they have been shot out of a cannon and have a winged cap, which as everybody knows, in the videogame world, allows one to fly? Further still, if we examine titles such as *Bishi Bashi Special* or *Super Mario Party*, we find that they offer a variety of stages so different from one another that they could be considered different games in their own right. Essentially, *Super Mario Party* and *Bishi Bashi Special* can be seen as collections of sub-games and the number, diversity and scope of these sub-games is a key marketing feature; '85 games jam-packed onto one bumper disc' proclaims the cover of Konami's *Bishi Bashi Special*. At least part of the point of these titles is to be found in the variety of these sub-games and, more specifically, in the variety of different techniques, skills and expertise they demand. Typically, each mini-game lasts only a short time, often considerably less than a minute and players are pitted against each other in head-to-head competition. As such, the object of the exercise is to apprehend the simulation, ascertain the rules of engagement, and master the interface as quickly as possible and before the opponent.

The wide variety of mini-games and the randomness of both their selection and, in many cases, contents is important as this goes some way to inhibiting the development of superiority through practice. Here then, the chance elements of game selection and initial state (*Alea*) can be seen to facilitate competition (*Agon*). Interestingly, and particularly the case with *Bishi Bashi Special*, the interface is presented as a constituent of the puzzle. Where videogame design typically speaks of the desire to make the interface appear as transparent as possible so as not to intrude into the player's experience and remind them of the 'game' (see Rouse 2001; Murray 1997; Laurel 1991 on the significance of 'immersion'), *Bishi Bashi Special* explicitly demands a conscious consideration of the location of buttons and switches on the control pad and, by utilizing disorienting graphical displays that signal the need for particular sequences of inputs but that require reprocessing by the player to translate them into appropriate motor actions, the game aims to make the controller feel both uncomfortable and a barrier to success (see Chapter 8 for more on engagement with the interface). In this way, through a combination of the barrage of demands placed upon the player, the variety of the mini-games (effectively 'levels' or 'stages') and the deliberate abstraction and problematization of the interface, the game can be seen to replicate something of the dizziness and disorientation of Calliois' (2001) *Ilinx* without recourse to the literal kinaesthetics of *R360* or the bemani ('beatmania') dancemats we noted in Chapter 2. As such, even though the game does not physically throw the player around in an aeroplane cockpit or require that they literally dance on a mock stage, some approximation of the sensation is delivered by rendering the interface uncomfortable and inappropriate.

It follows that, as they invariably offer different types of gameplay and demand variegated modes of engagement, videogames do not offer a consistent experience of interactivity. This simple but essential fact is frequently left unconsidered by commentators keen to proclaim the form as one oriented around being, doing, participation and (inter)activity and distinct from other media forms such as film or television (see Crawford 1984, for example). Moreover, the industry itself proclaims itself in the business of creating 'interactive entertainment' or 'interactive fiction' (see Aarseth 1997). It is useful, however, to ask what may appear at first a somewhat nonsensical question. How interactive is an 'interactive videogame'? It is initially tempting to tackle the question comparatively. That is, is playing a videogame more or less of an 'interactive' experience than watching television, or using the world wide web? This largely fruitless endeavour tells us very little about either activity and serves only to distance the forms and experiences. More usefully, we can reinterpret the question: how much time does a player spend actually 'interacting' with an 'interactive videogame'? This might still appear nonsensical and we may be tempted to present figures on patterns and durations of play such as those in Chapter 4, for example. However, it is possible to suggest a different answer:

Q: How much time does a player spend 'interacting' with an 'interactive videogame'?
A: Some of the time.

Examining a videogame such as *StarFox 64* (also known as *Lylat Wars* in some territories), we may begin to get some sense of the amount of 'non-interaction' in 'interactive videogames'. Upon placing the cartridge into the Nintendo 64 console and flicking the power switch, nearly two minutes of introductory 'movie' sequence precede the title screen with its pulsating 'Press Start Button' graphic that implores the player to initiate 'the game'. However, pressing the button does not plunge the player into the midst of frenetic space combat as might be expected, but rather initiates a further three minutes of pre-rendered movie sequence in which interactive control and the ability to influence proceedings is entirely wrested. Only once these sequences have run their course is control handed over to the player who can pilot their 'Arwing' spacecraft through the gameworld. Lest we should consider that these sequences are merely introductions, setting the scene, building tension, and creating atmosphere prior to the commencement of play which then continues uninterrupted until the objective is completed or all lives are exhausted, we need only consider a game such as *Metal Gear Solid 2*. Here, just as in *StarFox 64*, interactivity is frequently punctuated by non-interactive movie sequences or 'cut-scenes' as they are commonly termed.

It is important to note that cut-scenes are by no means the only sequences of non-interactivity in videogames. Consider *Pac-Man*. Certainly there are no lavishly pre-rendered movie sequences here, yet the game does not present consistent flow of play. Upon completing each maze, there is a break; the maze outline flashes white and, at certain points, there is a short animation of *Pac-Man* either chasing or being chased by the 'enemy' ghosts. A short sequence without doubt, but a punctuation in play nonetheless. Similarly, in *Gran Turismo 3*, after each race, the player is treated to, among other things, a table showing the race results and an action replay of the race. It is notable also that the game is portioned into tournaments, rounds, races and so on and the player does not simply start driving until they have had enough and turn off the console (though it should be noted that by selecting the 'free run' mode in which the rules and AI opponent cars that usually generate the agonism of the race are replaced by unlimited time and the freedom of paidea, the player has the liberty to do just this).

It should be clear already that 'playing' a videogame such as *StarFox 64*, *Metal Gear Solid 2*, *Gran Turismo* or *Pac-Man* involves often protracted periods that, prima facie, do not appear to bear the traits of 'play' as outlined in Chapter 2. It is worth recalling Rouse's (2001: 17) comments, 'players expect to do, not to watch'. There is in this proclamation a sense in which cut-scenes, or indeed any sequence of non-interactivity, is the necessary enemy of the videogame. This stance, as we shall explore in later chapters, appears to be grounded in the belief that non-videogame media are merely 'passive' (see Crawford 1984, for example) and that, as a consequence, the 'player' becomes 'viewer'. Clearly, such a position takes no account of the activity of the audience as espoused by media researchers for over two decades (see Morley 1992, 1986, 1980; Radway 1984, for example). Moreover, it overlooks the variety of purposes that such sequences serve. In order to frame the discussion of interactivity/activity that follows in Chapter 5, it is useful here to examine some of the structural commonalities of videogames and, particularly, to explore the functionality of what we might intuitively consider counterproductive breaks between 'play'.

LEVEL DIFFERENTIATION

Games as diverse in content and action as *R-Type*, *House of the Dead 2*, *Donkey Kong Country*, *Rez*, *Metal Gear Solid* and *Virtua Fighter 4* share one of the most common structural devices: the Boss stage. The Boss is an extremely common character or mechanism in videogames. It is essentially an end of level, or sometimes inter-level, guardian that must be defeated in order to progress to the next level. Bosses may be located at the end of every level, or as in the case of the *Super Mario* or *Sonic the Hedgehog* series, may occur at the culmination of a series of levels. Irrespective of how frequently they appear, they share many of the same characteristics and perform the same basic function. Where other adversaries encountered during the course of the standard levels may take relatively little time, effort or energy to overcome – though there are many more of them to compensate – the Boss is a more resilient foe taking considerably more skill and time to defeat. Typically, defeating the Boss is not a simple matter of repeating the basic attacks that may have disposed of the preceding minions. Bosses usually require a more complex strategy that accounts for their unique attack and defence patterns and their particular weaknesses. Very often, simply attacking a Boss by unleashing as much firepower as the player can muster can be counterproductive, as the Boss may absorb the energy of the player's attack or deflect it back at the player as in *Harry Potter and the Philosopher's Stone*. It follows that Boss encounters are frequently the hardest sequences in games, requiring not only dexterity and agility but also a tactical and strategic understanding of the Boss and the player's own capabilities and potentials. Essentially, the Boss may be considered as a puzzle. Even if the body of the level privileges exploration and offers a diversity of experience, the Boss stage poses a singular problem, pitting the player head-to-head with one puzzle. Typically, as with the *Metroid Fusion* example on p. 73, Bosses are puzzles with only a single solution.

Only one combination of attacks will be effective. Moreover, the player must learn the attack patterns of the Boss and utilize the arena in which the battle takes place to evade them.

It is notable that the Boss puzzle usually takes place in a separate and often completely self-contained space within the gameworld. In *Yoshi's Island*, for example (see p. 80), the space that precedes the Boss is blocked by a descending wall making evasion possible and forcing the player to engage with the puzzle. Effectively, the puzzle is made discrete from the remainder of the gameworld creating an inner 'magic circle' within which this particular episode is contained (Huizinga 1950). As such, set-piece Boss stages often represent the most memorable encounters of a given videogame as graphics tend to be somewhat more lavish than in the standard levels. Further reinforcing the significance of the encounter and the importance of the character, the Boss is typically physically larger and more imposing than the adversaries that came before, and is almost without exception apparently better equipped than the player.

Importantly, it is not left to visuals alone to convey the heightened level of intimidatory excitement that is to be manufactured. The videogame musician has a number of tricks. In *Donkey Kong Country*, there is a separate musical motif that accompanies each Boss. In addition to the arrival of this new and specific motif, there is an immediately noticeable change in tempo as the music speeds up to signal the increased significance of the encounter. Tempo, however, is perhaps the crudest of the musical tools available to the composer. Meter and key can also be manipulated. To add further to the perceived sense of freneticism and urgency, every few bars, the piece drops out of standard time and rushes into 7/8 thereby missing a beat. As this beat is missed the key of the piece rises a semitone. The effect is undeniably dramatic and creates a tension and urgency in the player. The shift into 7/8 pushes the action along giving the impression of time running out. Similarly, to the Western ear at least, such a time signature is relatively uncommon and breaks both the rhythm and melody in an unfamiliar and unnerving fashion.

The Boss represents the controlled climax to a particular sequence of gameplay, whether it be a stage as in *House of the Dead 2*, or a series of thematically and aesthetically related levels as in *Sonic the Hedgehog*. Indeed, many games make use of a mid-level mini-Boss to provide additional peaks of difficulty along the steady ramp towards the end-of-level Boss. Ultimately, a Boss or perhaps even a parade of Bosses from the game as in *House of the Dead 2*, provide the climax to the whole game. The 'standard' and 'Boss' level structure provides an effective and relatively simple means of building progression and development into a game. In addition to presenting its own unique sets of challenges, each standard level can be viewed as leading to these showdowns – and all of the levels can be seen as leading toward the ultimate battle with the end of game, 'Boss of Bosses'. Encounters with Bosses are presented in such a way to make paidea difficult. The Boss is incessant and attacks are typically relentless forcing the player into direct confrontation with the rules and parameters of the simulation on its terms. To reinforce this, not only is the Boss spatially separated from the remainder of the game which, in offering an often more open and flexible space facilitates the operation of paidea play, but also the player is usually unable to pause the action during these climactic episodes. The puzzle cannot be evaded, avoided or interrupted without ceasing the game.

Even in games that do not explicitly present Bosses as discrete and identifiable characters in their own stage or level, we can note much of the functionality at play. Nintendo's 1980s' Game & Watch titles are illustrative. In *Parachute*, players find themselves in a small rowing boat. Above them, an incessant stream of parachutists leap from a helicopter that, like a magician's hat, is spacious beyond its appearance. Parachutists fall according to one of three trajectories and the player must position their (equally spatially problematic) boat into one of the three landing spots at the right time to pick up the parachutists. Successfully collecting a parachutist earns the player a point. Failure to do so sees the parachutist chased through the water by a shark, and one of the player's three chances exhausted. The game is typical of what Rouse (2001) refers to as 'classic arcade games' in that it has an infinite structure. In fact, it is more accurately cyclical as the score counter resets back to zero after 999 parachutists are rescued. As such, there is no victory state to this puzzle, no 'solution'. The game ends either when the player exhausts their three chances or simply tires of the experience and leaves the parachutists to their fate. While, as in many other games, the ability to share and compare high scores represents a considerable incentive for play, the lack of an identifiable, externally-imposed victory state ensures that, for many players, the desire to 'zero the score', or even zero it a number of times, is an additional and important player-imposed ludus rule.

Interestingly, the lack of flexibility in *Parachute* and the incessant descent of the parachutists means that, unlike most other videogames, there is little scope for paidea. A cursory examination of the game suggests a simplistic structure with a steady ramping up of difficulty as the parachutists fall ever faster from the helicopter. The increased rapidity with which the parachutists emerge from the helicopter is not the only enemy of the player and *Alea* plays an important role. While the parachutists fall along one of only three possible trajectories, the sequence is random. However, with so few possible configurations, patterns soon emerge and players become primed to the appropriate combinations of button presses. However, similarly random is the parachutist who becomes entangled in a tree, swinging for an unspecified amount of time before continuing to descend towards the shark-infested waters thereby disrupting and further complicating the sequence. Moreover, the difficulty ramp is not so linear. Rather, the game is structured as a series of mini-crescendos. The rate at which parachutists fall from the sky steadily increases, and therefore the difficulty of the game steadily rises as players are required to perform the combinations of moves ever-faster. The tempo continues to rise until 100 points are scored whereupon the pace slows a little, only to begin another acceleration to the next 100 points. Through iteration of this process, the game gradually builds in pace and difficulty to the point where, like most 'classic arcade games', the player is finally defeated. However, importantly, while the game may appear on first inspection, to have been built around a very simple, linear speed/difficulty curve, the pacing of the experience is rather more sophisticated, building in peaks and troughs. The final 10–20 parachutists before each 100 sends the game and player into a dizzying blur, while the aftermath provides a welcome chance to compose oneself for the next onslaught. Thus, despite their being no explicit Boss, many of the functions such characters and stages perform can be seen in operation in *Parachute*. The Boss stage allows us to differentiate between two types of level within the game, and is one technique the designer can employ to build a difficulty curve, managing the tempo by manipulating the frequency of climactic encounters. However, the 'standard' levels that make up the majority of the game, between the Boss levels, need to be differentiated, also, so as to avoid the game becoming a monotonous parade through changing scenery.

Super Mario World 2: Yoshi's Island (re-released in 2002 as *Super Mario Advance 3* on GameBoy Advance) presents a superficially simplistic complicated gameworld. Compared with *Parachute*, *Yoshi's Island* employs lavish graphics and sound, and quite unlike *Parachute*, it appears to comprise a complex set of possibilities and tasks and offers the player a wide variety of capabilities and potentials. However, like most videogames, underpinning this complexity is a deceptively simple principal objective. Essentially, as Yoshi the dinosaur transporting Baby Mario on his or her back, the player is required to traverse a hostile landscape to the goal, which is quite simply the end of the level or course. Viewed in this way, it would be easy to assume that the player would simply become bored of the repetition. However, the game is not merely a linear journey from start to finish points. Rather, each level or specific region within a level is differentiated and it is these differences that generate the uniqueness of the encounter by presenting variegated challenges and puzzles for the player to engage with and apprehend and demanding the development of new tactics and strategies or the deployment of new skills. While level design is too broad a topic to tackle in intricate detail here (cf. Chen and Brown 2001; Pagán 2001a, 2001b; Warne 2001; Ryan 1999a, 1999b), it is useful to consider the way in which the theme and enemies affect the character of a level or sequence and contribute to the experience and pleasures of play. The theme of a level is most obviously expressed aesthetically with 'ice', or 'underwater' levels differentiated from those set on 'normal' ground by means of graphics, music and spot sound effects. However, most telling, the theme is expressed also through the game's 'handling'. For example, ice reduces friction and the player's character or vehicle slips and slides making accurate control more complex, while underwater levels simulate viscosity and buoyancy through the sluggishness of responses to inputs with eddies and undercurrents further hindering progress.

In narrative terms, the game may be seen as a series of what Brooks (1982) has termed *detours* and it is by differentiating levels that the retardations that Barthes (1974) also noted through the operation of the hermeneutic code are created and managed. At least part of the pleasure derived from gameplay, as Kinder (1991) has observed, emerges from the interplay between the desire to complete, to bring the game to its denouement, completing the level and delivering Baby Mario to safety, and the desire to prolong the encounter through *detours* and the activity of performance. As such, these retardations and delays are responsible for much of the potential pleasure of gameplay. We shall see in Chapter 7 that the gameworld is frequently constructed as a site ripe with potential for *detours*. Secret areas to discover abound and modes of play, whether inscribed within the game's own ruleset or superimposed as ludus rules by players, encourage thorough exploration in place of 'completion'. Play then becomes an act principally oriented around *detours* and the delaying of the 'end'. Foreshadowing the position of Fuller and Jenkins (1995) among others, the gameworld may be seen as a site rich in narrative potential, in which spatial stories (de Certeau 1984) may unfold through the transformations enacted by the player in their colonization of different levels and worlds. We shall return to this discussion later. Prior to this discussion, however, it is fruitful to explicate further the means by which the differences in levels are created. In doing so, we may also highlight the degrees of non-interaction constituent in videogame play.

Typically, each level brings a new set of enemies to confront the player often requiring different attack patterns for evasion or defeat. In *Yoshi's Island*, some enemies must be eaten, some hit with projectiles, and some popped like balloons (indeed some *are* balloons). Some require just one hit, others multiple blows while some are invincible and must be avoided. Similarly, while many wander the landscape following pre-defined pathways, seemingly unaware of the player's proximity, others actively seek out their quarry (see Pinter 2001; Stout 1997, for more on pathfinding principles). In *Yoshi's Island*, certain enemies even disguise themselves as benign objects or characters such as flowers only to reveal their true colours and attack. Consequently, confrontation with an enemy presents a challenge of the unknown – will it behave like an enemy previously encountered or exhibit new behaviours? As such, we note the continued significance of *Alea* (chance) in gameplay but also the degree to which the game requires (re)consideration and reprocessing of knowledge and experience gleaned from previous encounters in the game or perhaps with other games (as the flashing Boss feedback example on p. 73 indicates). This recalls Ricoeur's (1981) discussion of the continuity of consideration and the significance of memory and anticipation in the interpretation of narratives, and the work of active audience researchers who have pointed to the ways in which meaning is produced and texts are made to make sense by their users and readers (see Hermes 2002 for a succinct summary, also 1995). As we shall explore later in this volume, the need for such continuous scrutiny and re-evaluation of the experience and the application of the knowledge and technique deduced and acquired suggests that gameplay is not located solely within the present moment (what Juul 1999 has referred to as the 'now' of play). As such, the experience is cumulative, with the past read in the present and endings inferred from beginnings (see Chapter 6 for more on the activity of videogame audience).

BETWEEN LEVELS

While the flow and tempo of play within levels may be managed by various means (including automatic scrolling that ensures the passage through the gameworld is placed under the control of the designer), it is important to remember that, with notable exceptions such as *Parachute* with its implicit 100-point sections and continuous, unbroken gameplay, the various levels of a videogame do not usually follow on seamlessly from one another. Rather, levels are separated by breaks or intermissions. These may be as simple and short as text-only level numbers or names, or may comprise many minutes of lavish audio-visual spectacle in the form of cut-scenes. It is reasonable to ask why the play sequences of videogames need to be interrupted in this manner and many designers have implicitly signalled their dislike of them by attempting to create games with multiple levels differentiated as we have seen, yet offering the player the seamlessness and continuity of unbroken play that we have observed in Game & Watch *Parachute*. Two notable examples are *Spyro the Dragon* and *Half-Life*. In the former, levels are not announced with intermission screens. Rather, the loading of new level data is undertaken 'in the background' as Spyro leaps into the air. Certainly, there is a break from the action while the data loads, but the player never actually 'leaves' the gameworld. While *Half-Life* utilized the *Quake* graphics engine, the designers decided that by making the levels physically small enough, they could be loaded extremely quickly giving the impression of a far larger world even than a single *Quake* level.

There is a valorization of seamlessness in the videogame development community and designers lament loading times and technological limitations of even the most potent of today's PCs and consoles (see Rouse's 2001 motivations for play, for example). However, while many seconds or even minutes of loading screens may appear frustrating, this does not mean that seamless, continuous play is necessarily desirable. Clearly, the idea that seamlessness and continuity are desirable stems from the notion that videogame play must be about unbroken interaction. Periods of 'downtime' or non-interaction must, therefore, be detrimental to the overall experience of the game. However, while it remains true that they might be imposed to some degree by the technical limitations of gaming hardware, inter-level breaks play a variety of extremely important roles in managing and structuring the gaming experience. It is possible to identify a number of possible functions fulfilled by inter-level breaks. These functions are not mutually exclusive and particular games, or even particular moments in games, may utilize breaks to effect some or all of the functionality listed below:

- practical computing issues
- save point
- respite
- progress/feedback
- reward
- story development/exposition.

Even with today's powerful PCs and 'next generation' consoles such as PlayStation 2, Xbox and GameCube, technical constraints remain. While they may not be as restrictive as those of *Spacewar*'s host PDP-1, the constraints placed on the game designer by, for example, the finite amounts of RAM available effectively prohibit huge, seamless levels in favour of smaller, more discrete units. In short, because computer memory is required to store architecture and graphics as well as the rest of the game program, and because that computer memory is finite, games are necessarily split into portions that can be loaded and stored in the console or PC's memory. In addition, the comparatively slow access, retrieval and transfer rates of storage devices such as CD-ROM and even DVD-based devices used in most modern PCs and consoles mean that there is a 'downtime' during loading. Despite marketing hyperbole to the contrary (see Norman 1998), technical limitations limit the creative freedom of videogame designers to just as great a degree as in any medium. The fact that levels are restricted in size by the available memory of a host system or that the aesthetic, architectural or AI complexity of a gameworld is contingent upon the processing power of a series of microchips is revealing. Just as in filmmaking, technical (and institutional) restraints bear heavily on production thereby problematizing the designation of the designer as auteur.

SAVE-DIE-RESTART: MAINTAINING CHALLENGE IN MULTI-SESSION GAMES

It is a simple fact that videogames are not intended to be completed in single sittings. We have noted already that, not only are videogames frequently designed to be played in a variety of ways thereby giving rise to multiple levels of 'completion', but also that players may define their own games within the frameworks offered to them, even indulging in free paidea with scant regard for completion or outcome. However the game is tackled, it is clear that the majority of games present more than could be reasonably attempted, let alone 'completed', in a single session. The portioning of games into sequences or levels aids multisession gameplay. Even if the game does not offer a formal 'save point' at level boundaries, the arrangement of the game as a series of episodes facilitates both disengagement and return to play. It is convenient to leave the game at the end of a level as it provides a natural break and the player can return and pick up their encounter at the point they left. However, given the requirement to bring to bear the knowledge and experience of previous levels, disengagement from the game is potentially problematic and it is perhaps unsurprising to note that play sessions frequently last longer than intended. This may be due not only to the player becoming absorbed by the game as a result of its reinforcement schedules, both rewarding and frustrating players (see Loftus and Loftus 1983) but also may signal an altogether more conscious decision to keep playing. Usually this decision will be made so as to circumvent the possibility of forgetting valuable information as very few videogames offer any facility to recap progress upon re-engagement (see Chapter 4 for more on the duration of play sessions). It is notable that certain games prohibit this kind of access to later levels. Typically, coin-op titles (or home titles converted from coin-op) require that the player commences each session from Level One. This is particularly true of 'classic' arcade games of the 1970s/80s such as Namco's *Pac-Man* or Atari's *Asteroids*, although more recent coin-ops such as Sega's *Daytona USA*, effectively offer access through the selection of 'Beginner', 'Intermediate', or 'Expert' settings, whereby initial levels can be skipped.

The save point has become so engrained in the minds of players and designers alike that it is often used, perhaps even abused:

> 'I've got a great trap!' he told me gleefully. 'The player steps on this platform, it descends into a chamber he thinks is full of treasure, then a ring of flamethrowers go off and he's toasted.'
>
> 'What if I jump off the platform before it gets to the bottom?'
>
> I meant it as a solution, but he saw it as a loophole. 'Yeah, we'll have to make it a teleporter instead. You get flamed as soon as you materialize.'
>
> 'But, I mean what is the solution?'
>
> 'There isn't one!' He was astonished at me. 'I'm saying it's just a killer trap. It'll be *fun*.'
>
> 'So, there's no clue before you teleport that this might not be a good idea? Charred remains on the teleporter pad or something?'
>
> 'Nah, of course not. That's what the Save feature is for.'
>
> (Rollings and Morris 2001: 81)

Chris Crawford is similarly dismayed by this kind of gameplay mechanism and the reliance on the save point to backtrack. For one, it goes against one of the core videogame design principles that the player should not get stuck – there should always be a solution to every puzzle, predicament or problem presented to the player. Perhaps the single most frustrating experience that any videogame player can imagine is being stuck down a pit and not being able to get out. The only 'solution' is to reset the game and restart from the last save point.

For Frasca (2000), this save-try-fail-restart sequence ensures that videogames can never be considered 'serious':

Whatever you do in a game is trivial, because you can always play again and do the exact opposite . . . What the player does is experimenting rather than acting: she is free to explore any 'what if' scenario without taking any real chance. The problem is that usually 'serious' cultural products are essentially based in the impossibility of doing such a thing in real life. Hamlet's dilemma would be irrelevant in a videogame, simply because he would be able 'to be' and 'not to be'.

(Frasca 2000: 3–4)

The save point's tendency to trivialize player's choices is significant and, according to Maroney, this undermines the functioning of the videogame as a game as the implications of choices are not felt by the player. However, it is interesting to note the frequency with which walkthroughs and cheat-modes are employed by players as these would appear also to erode the challenge presented by the game, yet walkthroughs are sold alongside games at retail and abound on the world wide web, while cheat-modes offering infinite energy or the ability to see or even walk through walls are built into many games (see Chapter 7 for more on walkthroughs).

It is worth noting also that, for Loftus and Loftus (1983), part of the pleasure of videogames may be found in the ability to minimize regret. For Kahneman and Tversky (1982) a means of exploring regret may be found in the creation of 'alternative worlds'. These imaginary scenarios map potentialities and are dominated by the logic of 'what if'. Kahneman and Tversky assert that the closer the alternative world to the real events, the greater the regret. For Loftus and Loftus, videogames offer the opportunity and virtual space in which to explore these different scenarios:

'If only I had put on the radiation suit,' you say to yourself, 'I wouldn't have died that horrible death in the radiation chamber.' And since the alternative world in which you put on the radiation suit is very close to the 'actual' world in which you didn't, regret is very high.

(Loftus and Loftus 1983: 32)

As the desire to explore these alternative worlds constitutes a major motivation for play for Loftus and Loftus, it follows that the save-try-fail-restart cycle is important in enabling rather than undermining the integrity of the videogame.

THE DURABILITY OF INTER-LEVEL BREAKS

As an indication of their importance, breaks may be employed even where they are not technically necessary. Why, however, would a designer deliberately break up the flow of their game? Sustaining the required levels of attention and concentration in the player over lengthy periods is potentially problematic and inter-level breaks provide a welcome respite. However, this comes at a cost to the player and these breaks throw the player out of any rhythm they may have established. In this way, inter-level breaks can be seen as just one of the many ways in which players are deliberately 'put off' or distracted. Other techniques occurring within the game include the constant taunting of the player by the race commentator in *Ridge Racer Revolution*, the steadily increasing heartbeat sound effect in *Space Invaders* or the shift to an unfamiliar time signature as discussed on p. 78 (see also Chapter 9 for discussion of players' talk). The volume, velocity and pace of the game may mean that reflection on progress through the game, or contemplation of particular choices is not always possible. As such, the inter-level break provides a time during which the player can assess their performance, lament missed opportunities, or congratulate themselves on a job well done.

We have noted that assimilating the knowledge gleaned from a game is critical to the continued success of the player and strategy and tactics may be modified accordingly. To aid this reflection, in addition to a bare 'level completed' message, many games provide useful statistics highlighting successes and deficiencies in the player's performance. Games such as *Super Mario Kart* or *Gran Turismo 3* provide obvious feedback in the form of lap times that can be compared with the player's own previous best or built-in tables of the 'all-time best' thereby ensuring that the critical component of competition is not merely referenced to one's own performance but is given an 'external' dimension. The competition, even in a single-player game played by a lone player is thereby not merely one's past performance but effectively with the performance of other, not present, players. Sega's *Rez* displays percentages of kills while *Doom* offers tantalizing feedback as to the numbers of secrets revealed, encouraging repeated play, stressing the premature climax of the play session, and the potential for further *detours*.

Thus, while it is true to say that these inter-level breaks allow the player to reflect on their performance and gain a sense of progression through the game, the feedback provided by the game invariably encourages them to replay the level even after this apparent 'completion' as the information relayed to the player stresses how incomplete the performance was, and how much of the game still remains to be explored and unveiled. A 75 per cent kill rate may be read as nothing other than a 25 per cent miss rate and signals the need to replay and improve performance. This reinforces the multi-layered nature of video-game completion that must encompass not only the player's own ludus rules but which must also recognize the multiplicity of ways in which the game itself encourages play and replay through its own terms of engage-ment. As in the example here, it is often via 'non-interactive' sequences that such information is communicated to the player. It follows that such sequences provide a virtual commentary on the player's performance and offer a space within which critical self-reflection may take place. In this way, the sequences are not divorced from those that they frame but rather provide a point at which the player's attention is directed away from the relentless march forward to the 'end of level' or 'next sequence'. Within these 'non-interactive' breaks, the player's attention is critically focused on past events and performance either to encourage replay or to indicate the reason for the particular situation in which the player now finds themselves. For example, the route through a partic-ular level may be related to the player's performance. At a crossroads, for example, the course of the game may be dependent upon the player's health, the number or type of items they possess, or their score. *House of the Dead 2* shows a pathway map through the level indicating each branch (though tantalizingly omitting to indicate the criteria by which these 'automatic' route choices are made). Such materials invite the player to scrutinize the possibilities and potentialities of the game and the ways in which they might modify their performance to explore presently unavailable or inaccessible areas or elements of the gameworld.

We have seen in *Gran Turismo 3*, *Super Mario Kart*, *Rez* and *Doom*, that the inter-level break between levels need not merely involve the player staring at a blank screen while data loads from disc. The breaks have an identifiable functionality and contain substantive and original content. The break as reward provides a further illustration. Completing a level, winning a bout, attaining a podium position, or capturing the flag all require effort on the part of the player. Aside from the satisfaction of knowing you did it (or being able to save it to a memory card or hard drive for posterity), many games provide materials that simply reward the player for their success. Such rewards typically take the form of an audio-visual spectacle (an animation or movie sequence, for example). Upon completion of specific levels, Universal's *Mr Do!* rewards the player with (brief) animations showing the main characters, much like Namco's *Pac-Man*, while Namco's more recent *Tekken* series rewards the player with brief movie sequences upon completion of an allocated number of rounds and the defeat of the final Boss character. While the player may treat materials delivered during inter-level breaks as simple rewards, it is unusual that the sequences are intended to function solely as such. For example, movie sequences are considered by many players as little more than 'eye candy' or intermissions (and they can often be 'fast-forwarded' through so as to get back to what is believed to be the game proper), yet they are designed to forward a story line and deliver expositional narrative. Even the movie sequences in the *Tekken* series are intended to develop and progress the backstories of the various characters available to play in the game. That the narratives are largely inexplicable and in some cases downright bizarre perhaps renders them more readily understandable as audio-visual reward.

Games such as the *Metal Gear Solid* or *Final Fantasy* series are perhaps better examples of the inter-level break as expositional narrative space. *Metal Gear Solid 2* presents what may be the current apotheosis of narrative break with some sequences running for many minutes. Clearly, this demands that we further question our notion of what a videogame is and what types of activity it engenders, given that so much time is spent engaged in what we might not instinctively understand to be participative 'play'. The need to modify and reuse the techniques and knowledge acquired as the game unfolds, and most importantly that knowledge and techniques may be gleaned from non-interactive sequences, ensures that active participation is required throughout. The player must be constantly attentive and is encouraged to process and reprocess revealed information. As such, and in stark contrast to those critics that decry the non-interactive sequence as potentially injurious to the integrity of play, these sequences may be seen to be integral. Both framing and providing continuity, these non-interactive sequences demand high degrees of player activity as strategy and meaning are worked and reworked. While it is an oversimplification to suggest that interactivity and activity can be neatly confined within the boundaries of 'levels' and 'breaks' respectively, it is clear that to equate non-interaction with inactivity is quite erroneous.

The punctuation introduced by the breaks between levels creates a critical, reflective space in which action and performance may be scrutinized. Indeed, various techniques are employed to telegraph the need for such reflection perhaps drawing attention to the inadequacy of the player's performance, to the incompleteness and cursory nature of their journey thus far, or to the reasons for the trajectory of their journey and its contingency on performance criteria. It is clear, then, that the player's memory of their experience and travails is called into play. While at least part of the function of the inter-level break may be to provide respite from the high volume and velocity of many videogames, they are frequently also deployed in order to impart essential information with which the player may better tackle events to come, or may better understand and interpret experiences past. Importantly, the breaks may be seen to operate as spaces in which critical reflection is explicitly encouraged and facilitated. It is possible to argue, therefore, that these punctuating, 'non-interactive' sequences, the critical, reflective spaces they create, and the inherently participative activity that occurs as a result, encourage the player to construct and process their experience as a narrative. During such sequences, the player is explicitly required to search for causality, to read, decode and interpret the events of the game; to read the end in the beginning and the beginning in the end (cf. Cobley 2001a; Ricoeur 1981). The structure of the game, therefore, may be seen to influence or reinforce the way in which players (re)construct and make sense of their experience, positioning themselves as the central character within a personal journey or quest. In the next chapter, we will proceed to a discussion of the ways in which players may narrativize their experience and the debates surrounding the deployment of narrative theory in the study of videogames.

6

NARRATIVE AND PLAY, AUDIENCES AND PLAYERS
Approaches to the study of videogames

LUDOLOGY AND NARRATOLOGY

We have noted already that the premise of many videogames is reminiscent of Todorov's very basic narrative structure of (a) equilibrium disruption and (b) attempts to exact the inevitable resolution (note also Vogler's 1998 influential adaptation of Todorov's model). As such, the application of narrative theory to the study of videogames could be considered inevitable. However, it is fair to say that the issues of narrative – and, by inference, the audience for narrative – have caused considerable consternation in both the academic and practitioner videogame studies communities. Even though the discipline is in its infancy, a schism has already emerged between 'narratologists' and 'ludologists'. In fact, as we shall learn, the issue of narrative has, in some form or other, polarized almost all areas of the videogames community, from players to designers to academics.

In academic discourse, the question is frequently articulated through an analysis of the potential tensions between the activities of reading and interacting; and the tensions between dynamic, adaptive simulation and putatively static narratives. Among players and practitioners, the issue is more usually expressed through consideration of the cut-scene. In essence, while they seek to grapple with many of the same questions, it is possible to broadly differentiate academic narrative approaches from those of the player/practitioner communities. In the player/practitioner discourse, the player's desires appear to centre on an interrogation of the narrative 'elements' of the videogame (most notably the cut-scenes and introductions) and their relationship with the 'play' sequences of levels and interactivity, while within academic discourse, consideration is more usually of the videogame *as* narrative – that is, as reducible to narrative components. The key difference seems to be that the player/practitioner's *experience* of videogames demonstrates a sensitivity and awareness of the variegated nature of gaming that narrative theory can, at best, indicate and, at worst, completely neglect in its impulse to reduce videogames to mere narrative structures.

In this chapter, we will explore a range of approaches and responses to the issue of videogames *in* and *as* narrative and the ways in which such analyses indicate the value of a consideration of videogames within media and new media studies by drawing attention to issues such as audience/player (inter)activity and the dynamism and responsiveness of simulation, for example.

PLAYSTATION, CD-ROM AND THE CUT-SCENE

Jesper Juul (2001) has noted that one of the potential difficulties facing students of videogames is the selection of examples and case study materials. For Juul, the enormous variety of titles and types collectively assembled under the heading of 'videogames' means that critical approaches and theories are at risk of being unduly influenced by particular instances that may come to stand for the whole field. As with any academic approach to popular cultural artefacts, the sheer number of artefacts entails that, for better or worse, only generalizations can be made about their nature and use.

We might add to these concerns the observation that, through the operation and imposition of ludus rules, not every player plays a given game in the same way, nor do they necessarily seek the same pleasures from their play, as an examination of the use of walkthroughs and 'cheat-modes' indicates (see Chapters 7 and 9). However, it is more than just variety that renders the academic investigator's task problematic: while it is possible to identify underlying themes and constancies, it is also true to say that videogames have changed over time. This may appear self-evident, particularly if we consider the issue in terms, for example, of audio-visual sophistication. However, although the visual, and to a lesser extent audio, components of videogames are widely discussed within academic and popular discourse (see Crosby 2002b; Clark 2001a, 2001b, 2001c), there is more to the transformation of videogames than this.

In the previous chapter, we noted some widespread structural features that may be identified in videogames as apparently diverse as

Game & Watch, *Pac-Man* and *Metal Gear Solid 2*. However, while certain structures and forms might be seen to remain relatively constant, specific implementations have varied. For example, while the play sequence or 'level' framed and punctuated by a break or intermission has remained a staple of videogame design for many decades, the differences in the implementations found in, for example, *Pac-Man* and *Metal Gear Solid 2* are profound. Most notable among these differences is the relative balance of the play and break sequences, and since the launch of the PlayStation, videogame players have become used to increasingly extended sequences of 'non-interactivity'. Post-PlayStation, the 'cut-scene' has become part of the language of the videogame player and designer.

It is difficult to overstate the significance of the PlayStation in the development of videogames. Not only was the marketing of the system largely responsible for a shift in the cultural acceptance of the form and in the composition of the audience of players, but also, in bringing the storage and retrieval potentialities of CD-ROM to the console market, it can be seen to be responsible for one of the medium's most important aesthetic transformations. The PlayStation also incorporates an ability to play or 'stream' pre-recorded audio and video sequences directly from disc. It therefore included larger gameworlds with more varied levels; and it widened the scope for delivering extended introduction and inter-mission sequences. Pre-recorded ('pre-rendered') video clips, created 'offline', would be spliced between the play sequences of the game. Thus, the material could be aesthetically richer than that which it framed or introduced. Indeed, the ability to stream high-quality audio-visual sequences that has become part of the standard specification of all subse-quent consoles, has seen not only the elongation and continued aesthetic enhancement of these materials, but also their foregrounding in the design and marketing of contemporary videogames.

As we have noted in Chapter 3, these sequences may be created by specialized sub-teams of directors, lighting artists, musicians and cine-matographers, for example, who may not work on the playable levels of the game. One consequence of these protracted sequences to players of contemporary videogames such as *Metal Gear Solid 2* and *Final Fantasy X* is that a considerable amount of their time is spent engaged in activity we might not instinctively consider as 'playing'. In fact, to all intents and purposes, these sequences, along with the cinematic act of cutting, might easily be considered 'narratives'. Given this, it is perhaps unsurprising that the presence and role of cut-scenes has come under considerable scrutiny in both the player and practitioner communities with webrings, discussion boards and reviews home to criticism and commentary.

THE TROUBLE WITH CUT-SCENES: 'ACTIVE' AND 'PASSIVE', 'STORIES' AND 'INSTRUCTIONS'

Videogame designer Richard Rouse (2001) argues that the mode of current videogame engagement shifts from 'interactive, participatory play' to 'passive, detached watching'. Why would any designer, Rouse asks, go to the trouble of building an immersive world, an interface and controls that allow seamless access to that world, and labour over ensuring that engagement feels like second nature to the player? Why would they design and implement interesting, intricate and exciting puzzles, and then undermine it all by dragging the player out of the game by introducing what he considers to be mediating narration and 'non-participatory' watching? In asking these questions, Rouse also articulates what has been considered to be the main issue in understanding the videogame as a medium: whether play entails *interaction* and whether *narrative*, conversely, entails *non-interaction*.

For Henry Jenkins, the issue partly concerns the relevance of the materials presented in such sequences and their integration with the objectives of play sequences. Pointing to the fairy-tale, 'rescue plot' that accompanies the majority of Nintendo's *Super Mario* titles, he notes that: 'Once immersed in playing, we don't really care whether we rescue Princess Toadstool or not; all that matters is staying alive long enough to move between levels' (Fuller and Jenkins 1995: 60).

The games in the *Super Mario* series are by no means the only examples Jenkins could have selected. Namco's PlayStation conversion of *Soulblade* similarly illustrates the comparative lack of importance of narrative that it presents through its cut-scenes. However, for Rouse, the tension between cut-scenes and gameplay sequences remains between the (inter)activity of play versus the passivity of narrative. Rouse's position is supported, then, by Crawford's (1984) stance on videogame play as distinct from the 'passivity' of 'non-interactive' media or by Loftus and Loftus who claim that 'When we watch a movie or read a book, we passively observe the fantasies. When we play a computer game, we actively participate in the fantasy world created by the game' (1983: 41).

THE (INTER)ACTIVE AUDIENCE

Taking a somewhat similar starting point to Rouse, Crawford and Loftus and Loftus, 'ludic' or play(er)-centred approaches to the study of videogames have attempted to foreground the activity of *play* in their analysis of games, gamers and gaming. In this regard, they *stand* in contrast to so-called 'narratological' approaches in videogame study where the game is seen to be positioned as a text to be read. For adherents of ludic approaches to the study of videogames, collectively known as 'ludologists', narratological strategies are problematic for a variety of reasons. Chief among them centres on the apparent concentration on the 'text', and specifically the text as a static entity from which meaning can be deduced. For ludologists, it makes no sense to talk of the videogame text, in part because it cannot be seen to be constituted without the activity and action of the player. It is players who breathe life into and make sense of videogames. However, while it may be true that narratology does indeed evince a tendency towards static analysis, there is considerably more to narrative analysis than narratology.

Most notably, as a consequence of the impact of media and communication theory and engagement with a range of new media such as hypertexts and the world wide web that present 'interactive' elements, implicit within contemporary narrative theory is the notion of the audience as 'active'. As such, approaches such as those of Rouse, Crawford and Loftus and Loftus, while not explicitly 'ludic' in their designation, are perhaps guilty of a desire to position videogames and play as uniquely different from other media use. They demonstrate a willingness to position non-interaction as 'passivity' whether this be in the use of other, non-videogame media, or non-interactive sequences within videogames. Moreover, ludic approaches that harbour sensitivity to the player and his/her activity are perhaps not so opposed to narrative analysis as they protest.

Roger Silverstone (1994: 142–143) notes that the 'discovery' of the audience as active consumers and constructors of media messages in the later decades of the twentieth century was itself an invocation of a 'literary certainty' that found its roots in the work of Katz and Lazarsfeld (1955) and Schramm *et al.* (1961). Yet, this view of 'activity' is not simply focused on an individual such as an isolated player; in fact, in order to understand videogame use the definition of a videogame 'player' needs to be unpacked. Though the idea that an 'audience' does not simply play might appear somewhat contradictory, it is essential to note that videogame experiences are frequently shared by groups, perhaps crowded around a television set in a domestic setting or, as Saxe (1994) has observed, around coin-op machines in arcades (see also Green *et al.* 1998 and Chapter 9). Moreover, 'non-controlling' roles are commonplace. Map-reading 'co-pilot' or 'lookout' secondary player roles are frequently adopted by videogame users. All of these roles may seem 'passive', and it is clear that the ability to exert control over the game is different from that enjoyed by the primary player, as is the engagement and relationship with the gameworld (see Newman 2002a and also Chapter 8). Yet the phenomenon indicates that 'interactivity' is not simply a matter of individual players being 'active'. In this way, and perhaps echoing the sentiment at the heart of a ludic approach to videogames, 'playing a videogame' may be no more meaningful a phrase than 'watching TV' which, as Ang notes, 'is no more than a shorthand label for a variety of multidimensional behaviours and experiences implicated in the practice of television consumption' (1996: 68).

The nature of play and interaction with videogames has necessitated ethnographic research in a fashion that is not out of kilter with approaches adopted for the study of television (for example, Morley and Silverstone 1990). This is because, like television, the videogame cannot be considered as a technology or medium used solely on an individual basis outside any kind of context for its use. As Ang argues (1996: 69–70), the webs of intersubjective relationships within which media texts reside not only highlight the difficulty of understanding or predicting meaning but also problematize the notion of a medium's intrinsic or fixed potentials. However, even if we stop short of this radical contextualism, the importance of revealing the complexity and embeddedness of videogame consumption within everyday life, and in the context of other non-videogame consumption, is brought into sharp relief. By investigating only the experience and engagement of the primary player, the richness and diversity of uses of videogames is lost and claims to contextual sensitivity are hard to sustain.

A more nuanced way of understanding the videogame as a facilitator of interactivity is offered through the exploration of the practice of textual poaching and the exploratory nomadism of media fandom (itself a concept 'poached' from de Certeau 1984). The idea of 'poaching' suggests that all narratives, whether (self) proclaimed 'interactive' or otherwise, may be greeted by participatory, active audiences. As Jenkins asserts with reference to fans, they are far from being fascinated, awestruck apologists for the text. Rather, theirs is an exploratory role of revelation and discovery, a journey, as de Certeau would have it, that assimilates and reprocesses the dual responses of adoration and frustration. Fans poach from texts what best suits their own predilections. As texts may often antagonize and fail to satisfy, the fan might even be engaged in a 'struggle' with them. As Brooker (2002) notes in his investigation of the fan culture surrounding the *Star Wars* saga and, particularly, the mixed responses to the first 'prequel' film, *The Phantom*

Menace, such struggles bear results. They have culminated in, among other things, a re-edited 'fan cut' of the film with re-voiced and re-oriented characters.

Clearly, the examples discussed by Brooker take audience activity to greater length than is usually the case in the consumption of narratives. However, it is possible that these examples demonstrate important features of the phenomenon of narrative, features which are especially pertinent to those aspects of the videogame that seem to display a narrative orientation. In order to understand this point more fully and illustrate its application to videogames it is necessary to be clear about the distinction in narrative theory between 'narrative' and 'story', and particularly the work narrative engenders in the reader. As Cobley (2001a) has noted, the delineation of story, narrative and plot is foundational to any critical investigation of the area. For Bordwell and Thompson (1997: 66), story refers to all the events in a narrative whether explicitly presented or inferred by the viewer, while for Cobley (2001a: 239) plot refers to the causation that, by indicating a linkage between various story events, provides the justification for their depiction in relation to one another. In a sense, a story can seem to happen 'on its own', without too much help from a reader; but the chain of causation constituting plot requires a significant amount of work.

Importantly, and as a further challenge to Rouse's assumed passivity of narrative use, Ricoeur (1981) sees as fundamental to narrative a process of anticipation and recollection. Narrative, in this way, is a monumental effort to maintain time. Through the enacting of what he terms 'muthos' or 'emplotment', Ricoeur notes that temporality in narrative is more complex than the commonsense conception of instances arranged linearly on a 'timeline' would suggest and thereby the temporality encountered in narrative demands more than just attentiveness to individual events. Rather, successive actions are apprehended in relation to one another whether this takes the form of anticipated conclusions or backward glances to precipitant actions upon reaching the conclusion. As Cobley (2001a: 19) notes of Ricoeur, '[narrative] . . . is most importantly about "expectation" and "memory": reading the end in the beginning and reading the beginning in the end'.

It is not just play, then, which is a matter of interactivity in videogames. Narrative sequences can necessitate their own level of interactivity, requiring a certain degree of commitment from the player. It is true that this interactivity differs in quality from that of play activity itself, but it is a grave error to characterize it simply and in a contradictory fashion as an incitement to 'passivity'. Furthermore, this must be considered to be the case not only when narrative elements are bound up in the playing activity but also on the occasion of cut-scenes. This is not evident to all theorists of videogames, however; for Rouse, the 'narrated designer's story' is principally something that interrupts the business of play. He depicts it as a 'non-participatory' mode of engagement and an intrusive narration that reminds the player of the 'unreality' of his/her pursuit. And, in spite of what we have said about the activity involved in reading narrative, it seems that Rouse has a point. Many cut-scenes do not simply furnish a story element but appear to actually *undermine* engagement with the gameworld. The device not only seems to compromise the quality of the experience of play, then, but also, in the terms of this argument, seems to induce passivity. Indeed, some cut-scenes are quite explicit in presenting the endeavours of the player as 'unreal' gameplay. Let us interrogate cut-scenes further to determine how far these claims are borne out.

THE FUNCTION OF CUT-SCENES

Players and commentators tend to read and encounter cut-scenes as primarily narrative episodes concerned with exposition and causation, for example. This has been encouraged by the increased prevalence and high profile of videogames such as *Metal Gear Solid 2*, the *Final Fantasy* series and *Halo*. Each has been marketed largely on the strength of their storylines and the opportunities they offer players to step inside their narrative spaces, that is, to become more interactive. Yet, each has employed narrative devices which seem non-interactive, such as contextualizing scenes and cuts.

As we have indicated in our discussion of *Soulblade*, cut-scenes are not simply focused on breaking the (inter)action. Instead, they perform a number of roles, some of which are in tension and render various sequences problematic for the player. The opening, 'non-interactive movie' sequences of *Metal Gear Solid 2* establish scenario, location, atmosphere and the motivations of specific characters including those controlled by the player. However, in addition, like those sequences that continue to punctuate *Metal Gear Solid 2*, the opening serves to simultaneously explain (a) the overarching and immediate game and level objectives; (b) the mechanics of the game and rules of play; (c) the interface and function of buttons on the controller and objects within the gameworld; and (d) the mechanism for saving progress through the game. Thus, the cut-scenes serve to both advance the narrative in which players, as 'Solid Snake', find themselves embroiled while also imparting often practical guidance and instruction as to how to use the game and its associated technologies. This happens at a number of points throughout the game, perhaps to the extent that the signposting of gaming objectives may be seen to undermine the efficacy of the narrative reducing it to a series of gameplay elements.

This example might constitute inelegant scripting and overt 'infodumping' (see Roberts 2000); however, other tensions arise. The decision to relay information about the functionality of the game and interface via these cut-scenes presents something of a challenge to developer and player. Moreover, characters are seen to move inside and outside roles that position them both inside and outside the story world created by the events of the game. The characters are both participants in the narrative as it unfolds within the gameworld and participants within the activity of playing the game, imparting information about the operation of an interface of controls, memory cards, and access to the story or 'diegetic' world. Nintendo's *Luigi's Mansion* is another case in point: in addition to presenting gameplay advice in its cut-scenes, it sees the player's character nervously whistling the game's theme tune during play sequences, thereby further blurring the distinction between diegesis (story world) and non-diegesis (non-story world), 'inside' and 'outside' the gameworld. Similarly, if left unattended for more than a few seconds, Sonic the Hedgehog taps on the inside of the TV screen to wake the player!

In terms of other media this is not so unusual. So-called postmodern narrative or metafiction constantly reveals its own construction and points to diegetic and non-diegetic worlds (see Cobley 2001a: 71–100 and Waugh 1984). Indeed, metafiction employs devices that have been commonplace throughout 2,000 years of narrative's history. There is also an acknowledgement here of the reader's/player's activity in respect of narrative. Such activity has been underlined in the last 30 years by the commonplace use of technology by the contemporary television audience which through channel surfing, grazing, zapping, zipping and timeshifting, manifestly avoids immersion in individual narratives without abolishing the possibility of such immersion if it so wishes (see Ang 1996; Cubitt 1991). In the case of television, it is obvious that the act of zapping implies that there is some controlling consciousness of the narratives on offer – be it the television company or the narrator of narrative texts – that the viewer can choose to disregard by zapping to another channel. The introduction of remote and timeshifting technologies has not spelled the end of television narrative; nor have viewers lost their taste for immersion in television narrative.

So why is it that the videogame has come to be seen by certain commentators from both the academic and practitioner communities as incapable of sustaining its integrity in light of the revelation of the presence of a controlling consciousness – a narrator? Is it because of its status as game rather than narrative, hence requiring a different quality of immersion? If so, it is surely worth questioning whether other games require such a singular mode of engagement of their players and whether engagement with Monopoly, chess, or similar board games, for example, can be seen to wane if the player, even momentarily, 'steps outside' the act of play and the single-minded stance of 'the player'. Can the imposition of the 'Community Chest' or 'Go to Jail' be seen to equate with the disruptive intrusion of narration in a videogame? If so, we must surely question why this device does not undermine the board game.

Perhaps it is a matter of videogames sustaining a curious and sometimes seemingly antagonistic relation between narrative and play. Furthermore, it may be the case that the tension between the two arises from each wanting the other to do things which are, at present, only possibilities inherent in their modes.

NARRATIVE AND NEW MEDIA

It is certainly the case that the zipping, zapping and timeshifting engendered by remote controls and the VCR have brought the controlling consciousness of narrative into question by enabling the reader to manipulate those narratives that are available. Yet, it is the emergence of the digital computer that has really captured the imaginations of both theorists and practitioners concerned with interactivity. One only need look to science fiction for examples of the fascination with computerized narrative systems and the powerful effects that computerization can exert upon the form, construction and use of narrative media. The reason for this seems clear: computerization promises to enable narrative that is 'fully' interactive.

Star Trek's 'Holodeck' is a case in point and illustrates one of the central preoccupations for proponents of computer-enhanced narrative systems, of which the videogame might be considered one: the placement of the user, in a specific role, at the heart of the narrative. As showcased in *Star Trek: The Next Generation*, the Holodeck is a complex, computer-generated simulation environment that affords the crew members of the *Enterprise* the opportunity to take on a variety of narrative roles and 'step inside' the 'material world' of a story. Typically, though not exclusively, crew members are able to adopt the role of an already existent character – such as Sherlock Holmes, for example – whereupon they are inserted into a narrative space that plays out in real time. Importantly, the narrative virtual space of the Holodeck is populated by equally virtual characters that, as a consequence of their advanced artificial intelligence respond, again in real time, to the actions of the human user. The defining quality of the Holodeck, then, is its placement of the user/reader/audience (or 'interactor' after Ryan 2001) within the time-space of the narrative. On the Holodeck, there is no re-presentation of story elements, and the causation of plot emerges 'live' as a result of the performance of the interactor and their inter-actions with the computer-generated, virtual, characters. Although it exists only as a science fiction device, nonetheless, the Holodeck concept has been taken up by a number of theorists as an example of the way in which narrative could become, or how it may be articulated in three-dimensional, multisensory, 'virtual reality' environments and video-games (e.g. Murray 1997). Tellingly, initiatives in this area have also fed upon games for their inspiration: the 'Liquid Narrative Group', for example, seeks to utilize artificial intelligence algorithms for 'intelligent control of narrative' in order to produce 'novel, engaging and dynamic interactive stories . . . for interactive entertainment' using versions of *Unreal Tournament* among other games (Young and Riedl 2003; see also Cavazza *et al.* 2002; Young 2000).

Yet, if our understanding of the videogame is to be enhanced by its analogy with, and relation to, the computer's embodiment of narrative then the issue of authorship should be settled. Brenda Laurel (1991) addresses the matter by considering the computer as theatre, with the user in the role of participant rather than mere audience. Computer users not only join the actors on stage, but become actors, abolishing the notion of audience altogether. In Laurel's system, the computer program, the simulation at the heart of the application, effectively adopts an authorial role, albeit operating in real-time and responsive to the actions and activity of the performer. Thus, the system must be both adaptive and dynamic and exists to ensure that every action leads to the creation of a well-formed story (cf. Church 2000; Kreimeier 2000). Laurel's position is, therefore, a further elaboration of the convergence of narrative and play, where the roles of traditional figures such as authors as controlling consciousnesses have been transformed.

The attraction of the 'narrative you can play' has been considerable in the videogames industry and we have noted the marketing discourse that surrounds and frames such contemporary games as *Metal Gear Solid 2*, *Halo* and the *Final Fantasy* series. However, the idea of conflating narrative and play has a lengthy history in the videogames industry and can be traced at least as far as 'interactive fiction' such as *Myst*, *The Seventh Guest* and, especially, the text-only adventure games published by Infocom in the 1980s. Juul (1999: 15) draws attention to the company's early publicity claims of being able to place the player 'within the story', and generate credible characters with which to interact:

> We unleash the world's most powerful graphics technology. You'll never see Infocom's graphics on any computer screen . . . We draw our graphics from the limitless imagery of your imagination – a technology so powerful, it makes any picture that's ever come out of a screen look like graffiti by comparison . . . Through our prose, your imagination makes you part of our stories, in control of what you do and where you go – yet unable to predict or control the course of events.
>
> (Infocom 1983, cited in Juul 1999)

For many commentators on digital media, including Marie-Laure Ryan (2001), though, such Holodeck-style claims for digital narrative are a myth. Technologically, the type of immersive, 3D simulated environment demanded by the Holodeck is simply not within reach presently. Even 20 years since the Infocom titles discussed in the quote above, the adaptive, real-time 'narrative engine' remains science fiction. Furthermore, Ryan notes that current algorithmic sophistication makes it impossible for an AI agent to generate good plots in real time in response to the unpredictable actions of human participants (cf. 'Liquid Narrative Group').

According to Ryan, it is the relationship between interactor and character which is the key to understanding the extent to which full interactivity and immersion can take place. While it may be a defining quality of Holodeck narrative, first-hand participation is hugely problematic:

> Interactors would have to be out of their mind – literally and metaphorically – to want to submit themselves to the fate of a heroine who commits suicide as the result of a love affair turned bad, like Emma Bovary or Anna Karenina. Any attempt to turn empathy, which relies on mental simulation, into first-person, genuinely felt emotion, would in the vast majority of cases trespass the fragile boundary that separates pleasure and pain.
>
> (Ryan 2001: 6)

The implication, according to Ryan, is that these digital media narratives most likely lend themselves to the presentation of 'flat' characters with no emotional involvement in the plot. Such characters are defined not so much in terms of psychology but rather with reference to their capacity to explore worlds, solve problems, perform actions, and compete against enemies. This has important consequences for understanding the role of narrative in videogames. Put briefly, flat characters appear to be remarkably similar to the characters we have noted in videogames which are engaged with and understood as available capacities or sites for action (see also Chapter 8). And, as simple narrative 'functions', they also embody the characteristics that narratology has found interesting (see, for example, Propp 1968).

GAME TIME

As we have noted, a crucial factor in understanding narrative, especially as described by Ricoeur, is its enacting of time. However, a further layer of complication is added when narrative time is considered in relation with gameplay. Games designer and theorist Jesper Juul (1998: 3), relying on the work of the narratologist Genette (1982), suggests that classical narrative framework implies two distinct timeframes – the *story time*, denoting the time of the events told, and the *discourse time*, the time of the telling of the events. In addition, the *reading* or *viewing time* accounts for the fact that, even when an audience engages with a given narrative and the actors or characters can be seen 'here and now', the narrative still conveys a basic sense in which the events are not actually happening 'now'. Essentially, narratives are recounted and position their events in the past. Moreover, the plot imparts a causal logic upon the re-presented sequence – certain events necessarily lead to others. The game, however, according to Juul, exists solely in the 'now' and is defined by the possibility of influencing the game now. While in cutscenes, the distance between story time, discourse time and viewing time is reinstated in much the manner of narrative, the player's ability to act – or rather to play – implodes these relationships.

For Juul, it is quite simply impossible to have narration and interactivity at the same time:

> the game constructs the story time as *synchronous* with narrative time and reading/viewing time: the story is now. Now, not just in the sense that the viewer witnesses the events now, but in the sense that the events are happening now, and that what comes next is not yet determined.
>
> (Juul 2001)

According to this argument, shifts between narrated material and first-hand gameplay necessitate uneasy adjustments of the temporal distance between player and action. Videogame play is characterized by the player experiencing action in the gameworld at first hand in the immediate real-time of 'now'. Implicit in the positions of theorists like Juul as well as practitioners such as Rouse (2001) and Adams (2001a,

2001b), is the notion that mediating narration may be interpreted by players in much the same way as an obtrusive hardware or software interface might be experienced – as an unwanted barrier between the player and the experience of play.

If Juul's argument is to be accepted, then it has certain consequences for how we understand the 'contextualizing' narrative aspects of videogames which we considered earlier. The 'rupturing' effect of such narratives which reveal a controlling consciousness and which have been so prevalent in, but by no means limited to, other postmodern media, at least momentarily places the player 'outside' the game scenario. It interrupts the feeling of immersion. As we have seen, some feel that it appears to have no place in the videogame and cannot be sustained without the immersive experience being compromised. Certainly, the desirability of 'immersion' and the experiential dissolution of mediation has become a taken-for-granted trope in writings on technology (see Murray 1997, for example). Yet, such concentration on the creation of an almost mythic 'unmediated immersion', of course, suggests that the pleasure of videogames is located solely within the act of performance and play.

Perhaps the best way of approaching the character of the videogame, then, is through the distinction between the 'active' and 'interactive' audience. Where the active audience is defined in terms of the exploratory 'reaching out' into the text – guessing the twists and turns of the narrative, evaluating scenarios, testing hypotheses – that very reaching out is always already in tension with the transformative engagement – the interactive play – which actively renders the outcomes of the game uncertain. The causation in what Ricoeur (1981) refers to as emplotment in narrative could be seen to arise in the videogame not so much from narrative elements which are the focus of reader *activity* as from the act of real-time *interactive* play. In this sense, the act of play is seen not as an act of reading narrative but rather in producing – in a real sense – narrative *sequences as a consequence of* play.

However, one immediate objection must concern the structure and limited potentialities of videogames, and the danger of overstating the unspecifiable nature of the outcome. While they may be characterized by their adaptability, videogames do not present endlessly variable scenarios in response to player performance and the enacting of transformations on the simulation. The player's performance *is* bounded by rules, whether defined 'externally' by the game or imposed by the player. And, as we have also noted, a variety of institutional and economic factors mean that even the most apparently 'non-linear', branching games in fact comprise a finite number of levels. So, the videogame's 'adaptability' is a matter of a certain kind of bounded 'randomness'.

The game, *Blade Runner*, provides an excellent example of the parameters of randomness in the way that its simulation model enables the player to select non-playable character behaviours and responses. The designers of *Blade Runner*, in fact, have extended the concept by implementing a highly responsive, context-sensitive dynamic simulation model. The game offers the opportunity to play in the guise of either a replicant (a robot in the narrative) or a human. This mirrors the dilemma of the hero, Deckard, in the film upon which the game is based: how to distinguish between humans and replicants. Executive producer and lead designer, Louis Castle explains the operation of the simulation in an interview with Celia Pearce:

> If you play the game as if you were a *Blade Runner* human, it treats you like you're a human. So people perceive that at some point they've made a choice that puts them on one track or the other, which isn't the case at all. It's based on how you play the game, whether you hunt the replicants, whether you kill them, whether you let them go. Those are the things that give us clues as to what you think you are – and at any given point you can switch over.
>
> (Pearce 2002)

This kind of interactivity, then, is not only transformative of narrative actions but also changes the entire character of the game by being transformative of narrative functions. The fundamental choice in the game – replicant or human – might seem limited; but it has a profound bearing on subsequent choices and actions which illustrates the range of interactivity which bounded randomness facilitates.

Regardless of whether the outcomes of actions are predictable or are the product of complex calculations and the interplay of many irregular or arbitrary variables, the parallels between the constitution of videogames and narratives are now thrown into relief. All the possible elements that could be generated or presented by the game could be seen to constitute its 'story', while 'plot' or causation is created through the performance and activity of the player in dialogue with the simulation. This dialogue may negotiate a route through pre-defined scenarios as in games such as *Shenmue* or *Final Fantasy*. Alternatively, it may involve a more adaptive set of processes in which the performance is in interplay with combinations of AI or even randomness as in *Blade Runner*. Utimately, Frasca's (2001a) stance on videogame play may be seen as useful here. It explicates a process of narrative production in which the player and simulation are effectively engaged in the act of *emplotment*, charting a route through potential narrative space and engaging in digressions or *detours* (Brooks 1982). As with the desire that is enacted in narrative, the player's desire is to simultaneously progress and *retard* progress. A *session* of videogame play, then, can be seen as analogous to the reading of a narrative sequence: it assembles particular elements from the pool of possibilities whether finite or otherwise and affords them causation by creating and demonstrating their linkages. As we shall learn in the following chapter, for many commentators (e.g. Friedman 2002, 1995; Fuller and Jenkins 1995), the videogame may therefore be understood as presenting a (virtual) space through which the player journeys (the 'journey' being a traditional, staple action throughout the history of narrative). The narrative in the videogame can, therefore, be considered as machinery for the exploration of, and adventure in, cyberspace.

7

VIDEOGAMES, SPACE AND CYBERSPACE
Exploration, navigation and mastery

ADVENTURES IN SPACE

For commentators such as Aarseth (1997), space is a unifying theme of all videogames. The apparent containment of activity within a delineated 'magic circle' should not surprise as it is central to Huizinga's (1950) definition and is essential in separating the game from the 'real world' within which it is situated but from which it is, according to Huizinga, distinct (though it is important to note that scholarly discussions of cyber-spaces have pointed to the problematic nature of attempts to separate the 'real' from the 'virtual', see Farley 2000; Shields 1996). In this regard, the videogame world may be seen as analogous with the board or the table of (non-video) games such as chess, poker or roulette. Central to Aarseth's thesis is the observation that all videogames are intrinsically associated with the navigation and mastery of the spaces they present and produce:

> Every game of *Myth* is a fight for position in the landscape. To engage in battle without first securing a strong, ordered position, is in most cases to lose the game . . . units will go and do as ordered . . . but when the chaos of battle erupts, efficient control is no longer possible, and much therefore depends on how well the player has taken advantage of formation, land-scape variation, and knowledge of enemy positions.
>
> (Aarseth 1998: 11)

While in *Myth* Aarseth highlights the significance of spatial mastery through scrutiny, planning and the deployment of strategy, *Tetris* demonstrates the need for a real-time engagement with the spatiality of the gameworld. In *Tetris*, the player is charged with the duty of protecting their space by rotating and translating the descending geometric shapes that seek to overrun it. Similarly, in *Virtua Fighter*, the player is engaged in combat with both the opposing player and the arena within which the player must remain at all times. As such being 'knocked out' takes on a dualistic meaning and this spatial game rule can be seen to further separate the gameworld with a circle within a circle, or 'play-world' (after Newman 2001). Defending not only their virtua(l) body, but also their position in the ring, the *Virtua Fighter* bout can be seen as a tug-of-war in which exercising prowess brings the spatial dominance that equates with victory.

Typically, videogames create 'worlds', 'lands' or 'environments' for players to explore, traverse, conquer, and even dynamically manipulate and transform in some cases (the *Sim City* series is notable though by no means unique). As we have noted in the discussion of the typical structuring of the videogame into levels or stages, progress through a particular game is frequently presented to the player as progress through the world of the game. Perhaps the game reveals entirely new spaces, be they presented contiguously or as discrete areas physically unconnected with those already presented, or perhaps it encourages the re-examination of space already encountered. In this sense, gameplay may not only be seen as bounded in space, but also as a journey through it. Indeed, the marketing of videogames frequently makes explicit reference to spatiality particularly in regard to expansiveness and diversity. Nintendo's UK publicity for *Super Mario World* (re-released in 2002 as *Super Mario Advance 2* on GameBoy Advance) made great play of the expansiveness of its 96 different levels, each offering a unique challenge and comprising a vast world within which to adventure and explore. Here, the number of levels does not merely signal the potential longevity of the game as levels are neither encountered sequentially, nor is every level easily accessible, perhaps requiring considerable dexterity and even fortune.

Importantly, because all but the most skilful players are unlikely to even see let alone conquer all of the levels, spatiality is positioned as part of the puzzle and challenge of the game. Sony's 'Third Place' strategy publicizing the launch of PlayStation 2 made similar spatial reference pointing to the creation of elsewhere worlds. As such, we shall note on p. 113 that for many commentators (see Fuller and Jenkins 1995, for example), videogames may be seen to offer the equivalent of de Certeau's (1984) spatial stories, with gameworlds presenting sites imbued with narrative potential and in which play is at least partly an act of colonization and the enactment of transformations upon the space. Prior to this, it is fruitful to explore more of the structure and composition of the spaces and worlds produced within videogames.

VIDEOGAMES AND CYBERSPACE

As both Aarseth (1998) and Farley (2000) have noted, space is a common trope in contemporary media analysis particularly in relation to new media technologies, systems and experiences and nowhere is this more apparent than in the appropriation of the term 'cyberspace' from science fiction. Coined by William Gibson (1984) in his novel *Neuromancer*, though developed in *Virtual Light* (1992) and *Idoru* (1996) for example, and modified by a range of 'cyberfiction' authors (see Besher 1994; Stephenson 1992, for example), the term originally referred to the literally immaterial, intangible datascape created by, and accessible via, a network of computers:

> A graphic representation of data abstracted from the banks of every computer in the human system. Unthinkable complexity. Lines of light ranged in the nonspace of the mind, clusters and constellations of data.
>
> (Gibson 1984: 67)

However, despite the frequency of its deployment, the term 'cyberspace' is slippery and potentially problematic. The ambiguity arises not least because the term is applied in relation to a number of related but distinct qualities, characteristics and relationships that emerge from the proliferation and operation of ICTs (Information and Communication Technologies). For example, both the network topologies of the computers, routers and connections that comprise the Internet, and the environments created as a result on running applications on those computers are variously discussed as cyberspaces. As such, cyberspace may describe conceptual spaces within ICTs or the ICTs themselves (see Dodge and Kitchin 2001; Crang *et al.* 1999, for example). Moreover, while some cyberspaces such as VRML (Virtual Reality Modelling Language, often delivered across the world wide web) have explicit spatial referents and are designed to simulate 'geographic' space perhaps presenting 3D environments and employing Euclidean geometry, others may not (see Heim 1994; Sherman and Judkins 1992 for more on mid-1990s' visions of VR potentialities).

Despite the slippery nature of the term, cyberspace provides a useful frame within which to discuss the spaces of videogames. One important point concerns the 'immateriality' of these spaces that exist only insofar as they emerge from the code of a computer program. As such, they can be 'locationless' (see Adams 1992), though it is notable that the humanistic tradition in geography has long presented a conceptualization of places as centres of felt meaning and dislocated from the fixities of location (see Relph 1976 and note also Seamon and Mugerauer 1985; Tuan 1977). As cyberspaces are constructed in their entirety by 'space makers' (Holtzman 1994), it follows that, as Dodge and Kitchin (2001) note, their production is not merely a process of charting geometries but involves the definition of the properties of the spaces and their impact on the objects therein. As such, following Memarzia (1997), they note that properties such as gravity and friction do not exist in cyberspace unless they have been both designed and implemented. While the modelling of such forces is not in itself novel and players of *Spacewar* had to contend with inertia and the gravitational pull of heavenly bodies, the programming of 'physics engines' that govern the existence, magnitude and operation of such forces has become a recognizable and discrete area of work within the contemporary videogames industry. Moreover, it is frequently the case that the player is pitted against these modelled properties of the game space.

We have seen already that thematic level differentiation is expressed only partially through audio-visual means and that changes in the characteristics of the space produced and consumed within the particular level or sub-section play an important role. As such, it is the mathematical physics model of a car racing game in combination with the audio-visual rendering engine that differentiates 'ice' from 'tarmac', for example. Following Friedman (1995), it is again possible to suggest that the player is pitted against the (shifting) rules and parameters of the simulation. Furthermore, we note the significance of the 'feel' of the game and the kinaesthetic pleasures and challenges of spatial navigation. In this sense, and foreshadowing the discussion presented in Chapter 8, we note also that even in the absence of motion platforms and other paraphernalia (and excepting 'bemani', for example, with its considerable corporeal demands), the videogame may be seen to simulate, in the experience of the player, something akin to the kinetic pleasures of movement characteristic of Caillois' (2001) *Ilinx* class of games. It follows that, because the characteristics and properties of these constructed videogame cyberspaces do not, and indeed cannot, exist 'innately' as constituents of the gameworld but must be coded as elements in the simulation, there is no need for slavish correspondence to the 'laws' of offline, 'geographic' space.

In positioning virtual reality within a continuum of spatialities, Benedikt (1991) has noted the flexibility and mutability of space and time:

> the ancient worlds of magic, myth and legend to which cyberspace is heir, as well as the modern worlds of fantasy fiction, movies, and cartoons, are replete with violations of the logic of everyday space and time: disappearance, underworlds, phantoms, warp speed travel, mirrors and doors to alternate worlds, zero gravity . . .
>
> (Benedikt 1991: 128)

At least part of the pleasure to be derived from engagement with the cyberspace of virtual reality according to Benedikt, apparently comes from the ability to play with and within these elsewhere spaces replete with their uncommon, perhaps even unpredictable, spatial rules. We have obliquely noted already that videogame spatiality is not bound by contiguity and that tempo-spatial mutations are commonplace. Mario World's warp zones, *Turok*'s portals, or the reverse-gravity play modes of *Thrust* are cases in point. Discussion will return to the significance of these 'violations' (see pp. 122–123) with particular regard to the restriction of freedom of movement. However, preceding this analysis, it is useful to consider some further similarities between videogames and the cyberspaces described by scholars of the information society so as to further highlight the centrality of space in motivating players.

SPACES TO PLAY IN AND WITH

For some commentators, an important impact of Information and Communications Technologies (ICTs) such as the Internet has been the erosion of the significance of geographical distance and the effective shrinking of the globe (see Downey and McGuigan 1999; Cairncross 1997; Poster 1997; Castells 1996; Schroeder 1994). According to commentators such as Rheingold (1993), virtual communities form as much on the basis of shared interest (*gesellschaft*) as location or chances of proximity (*gemeinschaft*), though it is important to note Schuler's (1996) advocacy of 'community informatics' and the embedding of ICTs within local communities thereby reinstating the centrality of 'real' places and the local (see also Ackah and Newman 2003; Karim 2003; Wellman and Gulia 1999). It has been argued, by Rheingold and others, that a result of the perceived erosion of the significance of distance and increased spatial mobility has been a heightened sense of placelessness (see Relph 1976). With more than a nod towards McLuhan's (1964) 'global village', Rheingold sees applications of the same or similar ICTs as a means of offering an at least partial solution to these problems of dislocation and the fracture of community. As such, cyberspace may offer sites for refuge although Sardar (1995) among others is suspicious of the motivations of the cyberspace adherents wishing to harness the apparent perfectibility of digital media in order to create a simulacrum of community life. We shall return to the communicative, community cyberspaces facilitated through the web, for example, in Chapter 9 where we will note their importance in supporting and sustaining the participatory cultures of videogame fandom (Jenkins 1992; Lewis 1992). Here, however, we shall concentrate on the issue of cyberspace's offline surrogacy. Henry Jenkins, for example, has noted that videogames may be seen as a response to the restrictions of a contemporary urban existence in which shared, 'real-world' spaces of play are vanishing (Jenkins 2002: 59). Jenkins points to an increasingly restrictive urban environment in which the open spaces of the garden or backyard have become nostalgia for many:

> Watch children playing these games, their bodies bobbing and swaying to the on-screen action, and it's clear they are there – in the fantasy world . . . Perhaps my son finds in his video games what I found in the woods behind the school, on my bike whizzing down the hills of the suburban back streets, or settled into my treehouse during a thunderstorm with a good adventure novel – intensity of experience, escape from adult regulation; in short, 'complete freedom of movement'.
>
> (Jenkins 1998: 3)

Whether videogame worlds are truly seen by players to offer a substitute for real-world place spaces is, of course, a matter for debate (see Argyle and Shields 1996; Stone 1991; Turkle 1984) and it is possible to consider the ways in which other (non-video) games may offer similar opportunities and are affected by the same socio-cultural concerns and drivers. Certainly, an examination of the ways in which board games, for example, have developed and are used may provide support for such a stance. However, it is certainly true that the worlds of videogames, and the performance potential they promise, hold considerable appeal, particularly for those individuals whose physical environment limits their freedom of movement. Jenkins' position may help shed light on the popularity of videogames not only in the US and Europe but also in Japan where the interplay between urbanization and a restrictive physical environment creates enormous pressures on space. While film, TV, radio and literature all offer glimpses of elsewhere worlds, the opportunity to explore these spaces at first hand – to inhabit them, to get inside them, and it follows, to get away from the restrictions of the non-videogame world – can be seen as important motivations for play. In this way, the opportunity to adventure within these spaces, as we shall note below, is critical.

VIDEOGAMES AS SPATIAL STORIES

In a published dialogue between Henry Jenkins and English Renaissance scholar Mary Fuller (Fuller and Jenkins 1995), the centrality of space in the formation of the videogaming experience is seen to bear arresting similarities to the New World travel narratives of sixteenth- and seventeenth-century voyagers and explorers. A key feature of these narratives is that they are not driven or structured according to a plot or even through the development of characters as we might expect to find in 'traditional' or classical narrative forms. Rather, at the heart of these narratives is the transformation and mastery of geography – the colonization of space. In their discussion of these geographical narratives, Fuller and Jenkins employ Michel de Certeau's (1994, 1984) concept of the 'spatial story':

> For de Certeau . . . narrative involves the transformation of place into space (117–118). Places exist only in the abstract, as potential sites for narrative action, as locations that have not yet been colonized. Places constitute a 'stability' which must be disrupted in order for stories to unfold . . . Spaces, on the other hand, are places that have been acted upon, explored, colonized. Spaces become the locations of narrative events.
>
> (Fuller and Jenkins 1995: 66)

Thus, the transformation of place to space is a product of the action of narrative agents. Fuller and Jenkins illustrate the comparative imma-teriality of the devices of classical narrative in videogames (they talk of Nintendo only, though this can be read as a shorthand for videogames in the broader sense). The rescue plot that superficially frames the *Super Mario* series is just one example. For Fuller and Jenkins, the player is not engaged in a struggle to rescue the captive princess so much as they are engaged in a battle against the terrain of the landscape of the gameworld they have to traverse. What is really important to the player is staying alive long enough to get to the next level, and then the next, and the next . . . Staying alive long enough to explore, conquer and colonize the space of the gameworld – all 96 levels. Although Fuller and Jenkins do not explicitly turn their attentions to them, by examining popular 'God' games such as *Command and Conquer* or *Civilization*, we can see that the notion of colonization is not merely metaphorical, and just as in the case of many of the New World travel narratives, it is quite literal and frequently accompanied and reinforced by acts of aggression.

Interestingly, Fuller and Jenkins highlight a further characteristic of de Certeau's spatial stories that emphasizes their subjectified nature. De Certeau distinguishes between the 'map' and the 'tour'. Maps are objectified and speak of spatial relations. Tours, on the other hand, are personalized. A tour has a point of view, the point of view of the narrator. For de Certeau, tours describe journeys through spaces. They are accounts of spaces. Examining videogames as spatial stories, Fuller and Jenkins identify a dual pleasure. Part of the pleasure involves the transformation of geography itself, and part of the pleasure is in the subjectification and personalization of that transformation into first-hand experience. Thus, the player transforms place into space by taking over and conquering the gameworld. The unwelcoming, alien geography of the gameworld is settled. Moreover, the map becomes a tour as the player constructs a personalized account of the geographical transformation. In presenting and facilitating a personalized journey through space characterized by the tangled *detours* of exploration, videogame play may be seen to be located within a venerable Western narrative tradition that has produced such varied texts as *The Odyssey* and *Star Trek* and that, by virtue of the popularity of radio, television and literary series, for example, continues to be a dominant narrative theme.

Through his examination of *Civilization II*, Ted Friedman (2002) has taken Fuller and Jenkins' concept of the videogame as spatial story further still and has made a number of interesting observations, particularly with regard to the personalization of the 'tour' and the viewtype of the game. *Civilization* is a 'simulation' or 'God' game that places the player in the role of the leader of an empire-building nation:

> In Civilization II, you're responsible for directing the military, managing the economy, controlling development in every city of your domain, building Wonders of the World, and orchestrating scientific research (with the prescience to know the strategic benefits of each possible discovery, and to schedule accordingly). You make not just the big decisions, but the small ones, too, from deciding where each military unit should move on every turn to choosing which squares of the map grid to develop for resources.
>
> (Friedman 2002: 2)

For Friedman, it is abundantly clear that *Civilization* is principally concerned with the colonization and mastery of geography, 'gameplay in

Civilization II revolves around the continual transformation of place into space, as the blackness of the unknown gives way to specific terrain icons' (Friedman 2002: 4). However, while *Civilization* may be read as the archetypal spatial story, Friedman notes an important difference. In simulation games, he suggests, the map never becomes a tour. The viewpoint, the perspective of the player as they gaze onto the world never changes and remains resolutely fixed, overhead and detached. This is important because the apparent level of detachment suggests that the player cannot 'get inside' the map in the way de Certeau proposes. The tour, after all, is the individualized account of spatial experience (see also the 'walkthrough' on pp. 120–121). The narrative is structured around the personal experience of the space. For Friedman, the simulation game does not do this. Instead, it offers a different form of narrative altogether – the story of the map itself. 'The map is not merely the environment for the story; it's the hero of the story' (Friedman 2002: 5). For Fuller and Jenkins and Friedman, the centrality of space, and particularly the mastery of that space, is a defining feature of videogames. Whether, as Fuller and Jenkins' examination of Nintendo games suggest, the 'map' becomes 'tour' and the narrative is one of the traveller moving through the conquered spaces, or as Friedman's study of *Civilization II* indicates, the narrative is one centred around and about the map itself, it is clear that videogames, space and place are inexorably bound. To both Fuller and Jenkins and Friedman at least part of the pleasure of videogame play is derived from the transformation of place to space, the eradication of the unknown and the bringing of uncertain geographies under the control and influence of the player. In other words, videogame play involves the reinstatement of order that Danesi (2002) has identified as a key pleasure of puzzle-solving. In this sense, videogames may be seen, in part at least, as spatial puzzles.

SPATIAL TYPOLOGIES

Thus far, our discussion may have implied an unwarranted homogeneity of videogame spatiality. Videogames differ widely in their implementation of space, and the freedom of movement that they offer players. Both *Super Mario Sunshine* and *Ico* present significant differences, for example, in spatial contiguity with the former making use of warp zones and the latter presenting a model of a castle and its environs with the integrity of an architect's visualization. Aesthetically, too, videogame spaces vary considerably. Indeed, in the earliest days of gaming, non-graphical representation was commonplace and is still regularly encountered online in MUDs (Multi-User Domains), for example. Faced with such a variety of implementations, it is inevitable that theorists have attempted to design typologies so as to more neatly classify the forms and highlight the commonalities and dissimilarities in the various representations and uses of space.

Aarseth (1998) makes a number of useful distinctions that provide a basis for the differentiation of videogame spaces. He identifies three issues: the degree of integration of player-character and environment; the degree of manipulative control the player can exert on the landscape of the gameworld; and the degree of openness of spatial constructions ('indoor' and 'outdoor' games).

Man against the environment

For Aarseth, the approach of many early videogames can be described as the pitting of 'man against the environment'. The representation of the player and that of the world are distinct. The player's character is in the world but not necessarily of it. This perhaps helps us understand the hostility of game environments such as Mario World. The player, as Mario, is placed in this alien world and must defend themselves against its unknown dangers. *Ico* can be seen in a similar light. The player as *Ico*, is transported to the castle at the outset of the game. The space is not familiar and the player is at once immersed within it, yet pitted against it, viewing it as an abstracted puzzle. The player-character stands out from the landscape of the gameworld. Aarseth contrasts player-character uniqueness with games such as flight simulators, the most famous of which is doubtless *Microsoft Flight Simulator*, in which the player-representation is integrated in the gameworld (Aarseth 1998: 6).

Manipulation of the gameworld

It is clear that during the design process, the producers of videogames or 'space makers' (Holtzman 1994), have the space of the gameworld and its various properties under their direct and complete control. Level design is at least partly a process of agonizing over architecture and the placement of obstacles (see Beram 2001; Chen and Brown 2001; Johnson 2001; Pagán 2001a, 2001b; Warne 2001; Ryan 1999a, 1999b). However, the degree to which these crafted spaces can be subsequently influenced or manipulated by the player provides one way that the wide variety of game types can be distinguished. Aarseth draws a distinction between games such as *SimCity* and *Warcraft*, for example, that afford the player considerable control and influence over the gameworld, and games such as *Doom* and *Super Mario World* in which the player exerts little or no constructive influence. In *Doom* and *Super Mario World*, the game-world is far more static than in *SimCity* and *Warcraft*, where manipul-ability and spatial dynamism are important features of gameplay. Jaqueline Tivers (1996: 19) also notes the distinctiveness of landscape creation and manipulation in certain games. 'In *Populous*, the player creates the physical landscape by raising and lowering the land, forming areas of water, and producing earthquakes, volcanoes or swamps on opponents' land.' This chimes with the position of some cyberspace commentators who point to the infinite fluidity of virtual spaces. *In extremis*, Novak (1991) sees in the immateriality of online space a 'liquid architecture'.

While the differentiation between static and dynamic gameworlds is both important and useful, two issues must be raised. First, as Aarseth himself notes, the distinction between 'static' and 'dynamic' is not always easily made as boundaries are blurred in many games. Although ostensibly static, and offering the player none of *Populous'* manipulative influence over the game space, *Duke Nukem* nevertheless allows the player to smash windows and blow up trees, for example. While such activity does not constitute, or even contribute, to the stated game objective, it certainly helps generate a believable, immersive world of verisimilitude. In this way, it can be seen to perform an important function in estab-lishing and maintaining what Barthes (1992) refers to as the 'reality effect'. However, it is not true to say that such activity is trivial or plays no role in the player experience of the game space. Games such as *Virtua Cop* or *House of the Dead 2*, go further and the player is not only able but is encouraged to explore the space in which the principal action of the game takes place, and is rewarded for doing so. While the main object-ive of *House of the Dead 2* is to dispatch a seemingly endless horde of zombies, a number of rewards (additional weapons, extra health, for example) are hidden 'within' the environment (in dustbins or barrels or behind 'destructible', perhaps 'liquid', walls, see p. 120 for more on implied pathways through game spaces). *Metal Gear Solid 2*, too, rewards lateral thinking and the utilization of the game environment – hiding the stunned bodies of guards in lockers so they are not detected, for example. Such tactics encourage the player to consider the characters, space and action holistically rather than as an assemblage of constituent components, thereby apprehending the simulation in its entirety. Although it may be hostile, the gameworld is also the friend of the player. The player is not merely pitted against the game space but rather is encouraged to think about how the space can be utilized to assist in the attainment of the player's objective.

Indoor and outdoor games

It is relatively easy to identify the profound differences between an 'indoor' game, such as *Doom* that places the player in a closed labyrinth of narrow corridors, and the open landscapes of *SimCity*, *Myth* or *Command and Conquer*. However, the broad distinction between 'indoor' and 'outdoor' game spaces is something of a misnomer. In fact, the principal difference to be highlighted here does not concern the interior or exterior location of game spaces. Rather, it is the restrictiveness of those spaces that is key. A number of levels in *Doom* are set wholly or partially outdoors. Yet, even these exterior spaces are governed by the same basic spatial restrictions experienced indoors: 'Even in outdoors scenes in DOOM the landscape is riddled with obstacles and narrow paths. What may seem like a naturalistic world is in fact a constrictive topology of nodes and connections between them that interferes with unhindered movement' (Aarseth 1998: 7). Moreover, while *Myst* with its wide vistas and panoramic views seems an obvious 'outdoor' game, the labyrinthine discontinuity of its gameworld reveals that it has more in common with *Doom* than the (single-player) world of *Myth*. *Myst* comprises a network of still images linked together by web-like 'hotspots'. While each still image generates a naturalistic spatial integrity, the assemblage of them in series is discontinuous.

NAVIGATING CYBERSPACES

Given the centrality of space and spatiality in a variety of digital media, it is perhaps surprising to note the number of studies that highlight the problematic nature of virtual space particularly with regard to navigation and learning spatial relationships (see Witmer *et al.* 1996, for example). In their survey of extant research, Dodge and Kitchin (2001) have pointed to the visual simplicity of many virtual environments as a contributory factor. Certainly, technological constraints, especially in relation to available memory, give rise to worlds in which elements are recycled and repeated and selected from available 'libraries' of assets. Thus, forests of identical trees are not uncommon. Moreover, the rendering of (3D) objects in digitally generated space is frequently achieved through a process called texture mapping. Here, a small sample of a given texture (e.g. grass, stone, brick) often acquired from the 'real' world through scanning is applied to an object or surface. Where the texture sample is insufficient to cover the full extent of the surface, it is repeated or 'tiled'. As such, an expanse of virtual grass typically comprises a single texture sample repeated as many times as is appropriate (potentially *ad infinitum*), and therefore it follows that there is no potential for local variation. In such potentially homogeneous surroundings, landmarks provide one useful means of orientation and assist in the process of cognitive mapping (Siegel and White 1975). Implementations of a virtual sun, though, may prove somewhat more problematic. In *The Legend of Zelda: Ocarina of Time*, the gameworld (the land of 'Hyrule') is modelled replete with a sun and moon that move across the virtual heavens. However, the temporal compression presented within the game in which a day lasts only a few minutes renders it difficult, especially for the inexperienced player unfamiliar with the game's temporality, to utilize these celestial bodies as navigational aids. Moreover, time freezes when the player enters certain spaces such as shops or dungeons (compare also with the sequel *The Legend of Zelda: Majora's Mask* in which the entire game takes place over the course of just three (virtual) days).

It follows that many of the mechanisms utilized offline to navigate and conceptualize space by reference to its own properties and characteristics are either unavailable in cyberspace or are presently too lacking in richness to be greatly useful. One solution is the compass. Typically overlaid on the screen, in many implementations it dynamically rotates ensuring that North, for example, is always 'straight up' thereby lessening the cognitive load on the player engaged in frenetic activity. In many games, including *Zelda*, the compass is not provided as part of the standard 'equipment' but rather must be located within each level and while it is possible to play without it, navigation and orientation are potentially problematized. Indeed, its denial may constitute a player-imposed ludus rule.

Wayfinding and navigation is a crucial issue in videogaming and potentially distinguishes the forms from other cyberspaces. Though, in attempting to chart ways in which the legibility of cyberspaces may be improved in order to make wayfinding easier and more efficient Darken and Sibert's (1996) project is perhaps different from that of the videogame designer, who as we shall learn may seek to deliberately problematize pathfinding as part of the puzzle and challenge of the game, their findings are nonetheless interesting. Key among their suggestions for improving the usability and efficacy of virtual spaces are the inclusion of means of signposting the space, and the sub-dividing of environments into smaller spaces. It might seem that providing navigational cues to players of videogames would be detrimental to their experience and undermine the operation of the game given what we have noted in relation to the significance of player exploration. However, quite explicit signposting is often employed in videogame space. *Yoshi's Island*, for example, places directional markers along its various levels to help the player navigate the world. Or at least, they appear to help. In fact, such literal signposts are frequently used to misdirect players deliberately steering them into traps or away from specific items or areas. As such, the player is presented with a continual quandary as to whether or not to trust the signposting. Open, expansive spaces such as those presented in *Halo* nonetheless have implied pathways through them, yet the player remains always aware that to simply follow is to risk missing the pleasure of potential *detours* while to stray is to run the gauntlet of becoming lost and disoriented precisely for the reasons highlighted previously but also because the space is made deliberately treacherous and difficult to navigate (players are required to negotiate narrow cliff paths while plunged into near-total darkness, for example).

As we noted in Chapter 5, *Doom*'s 'secrets revealed' feedback speaks of unrealized potentiality and demands that the player consider what might have been and still could be on re-examination and exploration (see also Loftus and Loftus 1983: 30–33 on regret and reinforcement schedules). If signposting is used unconventionally, their second suggestion is more courteously embraced. We have noted already that videogames are, almost without exception, separated into different levels or stages and while this structure may have arisen from technological constraints, for various reasons, including the patterns of play and engagement, it remains durable and convenient. The net result is that the sprawling and potentially unfathomably complex gameworld of *Halo*, for example, is presented in more manageable, and more easily navigable and cognizable, chunks. However, despite this chunking, the process of cognitive mapping (see Lynch 1960) remains a complex task in the videogame given the deliberate convolution and maze/labyrinthine structures of many game spaces and other forms of cartographic representation that are frequently found within games. Indeed Darken and Sibert conclude that the provision of supplemental materials, not only detailing the spatial relationships within the virtual world but also identifying the user's location therein, are an essential tool in maximizing efficacy (see also Kitchin and Tate 1999; MacEachren 1995). In addition, commercially and fan-produced 'walkthroughs' further aid the player in their negotiations and dialogues with game spaces. Walkthroughs are texts outlining, in often-painstaking detail, the potentialities of the gameworld. More than commentaries on the game, walkthroughs serve at least two purposes (in addition, their production is a significant and visible mastery of the game, see Chapter 9). First, they frequently offer maps detailing the full extent of the gameworld including 'secret areas'. Second, they offer narra-tivized, egocentric accounts of the ways in which the player may tackle the game that present a relational space much like the pirate's treasure map (take ten paces forward, you will come to a rock, take three paces left . . .) that indicate the ways in which, for example, secret areas may be uncovered.

The implications of the use of walkthrough texts is considerable and most obviously the nature of the (spatial) puzzle is apparently undermined. Certainly, their use appears to signal a privileging of performance over puzzling and it must be noted that, for many games, there is considerable latitude for success or failure even once the objective or puzzle solution is known. Furthermore, walkthrough texts typically contain more than just the description of tactics and strategy and may provide an additional source of player-derived rules to be overlaid onto the simulation. Despite their prevalence both online and at retail where they are sold alongside the games they describe (and their presence here must surely be intended to communicate value for money through expansiveness and the virtual impossibility of extracting all of the potential pleasures of play) to date, scholarly investigations have overlooked them entirely and research is urgently required.

SPACE AND GAMEPLAY

The notion of naturalism in videogame spatiality is problematic. Examining games such as the *Super Mario* series provides clear evidence that non-contiguity is commonplace in videogame space. Even in MUDs where great care is often lavished to ensure that the various rooms that comprise the gameworld are topologically accurate – in that they constitute a space that could 'work' in the real world with explicit and contiguous spatial linkages and routes – the availability and use of teleporters undermines this correspondent realism. Because 'actually' traversing the rooms between the current and desired locations takes valuable time and effort, it is usual for MUD players to simply teleport by '@joining'. *Turok: Dinosaur Hunter* uses a similarly non-linear teleport system. An array of portals non-spatially link the central Hub to individual levels (similar to the paintings in *Super Mario 64*). In a style popularized through science fiction such as *Star Trek* (see Benedikt 1991), the player walks through the appropriate portal in the Hub whereupon they appear at their chosen level. In their discussion of *AlphaWorld*, Dodge and Kitchin (2001: 160) note that while the use of teleportation is expedient particularly given the expansiveness of the space, it is potentially counterproductive in discouraging exploration, limiting the potential for chance encounters and weakening the users' knowledge of spatial relations (see also Anders 1998).

That the outdoor spaces of *Doom* are structured as the corridors found inside is telling. Space here is manipulated and implemented by the game designers to enact a specific type of gameplay experience. In other words, spatial representation is subordinate to gameplay. Therefore, while videogames can be said to be spatial, it is their deviation from the patterns of 'real space' that enables them to function as games. Gameworlds are designed and constructed to offer specific kinds of experience and not to model or represent specific spaces. Experiences in gameplay are enabled by the manipulated space of the gameworld. The constrictive nature of *Doom*'s spaces is essential to create the tension of the action – the player is literally forced down certain routes into combat. Too high a degree of freedom of movement will simply allow the player to evade confrontation. It is important to note that, while videogames can be defined as a form by the emphasis they place on spatial exploration, navigation and mastery, the player's exploratory and navigational freedom is often severely limited in order that particular types of gameplay can be enacted. In this way, we note a potential tension between the operation of Barthes' (1992) 'reality effect' and the operation of the game. Indeed, as we shall learn, the overt restriction or constraint of spatial freedom may be negatively cited by players and reviewers against games.

The inability to traverse a hill with the same incline as one assailable previously, or the presence of 'invisible' walls that explicitly but inexplicably limit exploration, represent the clumsy imposition of devices that upset the verisimilitude of the experience and highlight the intrusion of the simulation. This is revealing as the dialogue between the player and the simulation is brought into focus again. One reading of the balance between verisimilitude and the operation of the simulation perhaps suggests that the simulation is not as central as theorists like Friedman might encourage us to think. However, it is important to note that the impositions and intrusions under consideration here represent not merely factors working to contradict the reality effect of the gameworld. Rather, these impositions restrict the operation and deployment of apparently reasonable player strategies. As such, they can be alternatively read as failures or limitations (or, by some players, more charitably, as perhaps just facets) of the game's simulation model. As such, the integrity of the simulation is not under threat from other, in this case spatial, factors, rather the integrity of the simulation itself is challenged by the player. In these cases, the limitation of the simulation model is expressed spatially.

Taking part of our lead from industry parlance, it is useful to distinguish between the spatial construction and spatial experience of gameworlds. We can briefly illustrate the importance of the distinction by considering a one-on-one combat game. While the arena and background scenery of the game may be presented in glorious three-dimensionality, the players are not granted anything like three-dimensional freedom of movement within it. It is usual in one-on-one combat games like *Soul Calibur* or *Tekken Tag Tournament* for combatants to be effectively placed 'on rails'. The two fighters move along a fixed axis that permanently orients them face-to-face. Players can sometimes employ a sidestep move, but this merely realigns the axis and the two characters remain facing each other. A roving camera that continuously moves about the arena displaying the bout from different angles usually masks the constraints of the engagement. It is reasonable to ask why the players' freedom of movement should be so restricted given the desire common to designers of 'virtual reality' environments (see Rheingold 1991, for example) and as espoused through popular discourse in the form of *Star Trek*'s Holodeck, for example (see Murray 1997 and Chapter 6) to offer a rich and free experience apparently mirroring and simulating the presumed full freedom of movement available in the real world.

It is noteworthy that many, especially early, experiments and commercial applications of virtual reality offered a freedom of movement that, in fact, significantly undermined their verisimilitude. The architectural walkthrough that places the user as a floating, disembodied, almost God-like entity gliding through (and in many cases, quite literally through, as doors and walls dissolve presenting no resistance) a building, while perhaps useful, falls some way short of the grand claims to offer an exacting, simulated reality (see Heim 1994; Sherman and Judkins 1992; Rheingold 1991). Videogames designers have learned that full 3D freedom of movement is potentially confusing and unwieldy. Importantly, these are *game*worlds with specific objectives and most importantly rules. As such, in a fighting game, it is important that the players fight. Allowing them to move away from opponents not only facilitates evasion but also makes it difficult to engage in combat even when intended as the task of 'lining up' players in three-dimensional space so they can hit each other is problematic (as experiments like SNK's *Samurai Spirits 64* have proved). Removing these considerations from the player, and limiting freedom of movement to 'back', 'forward' and 'jump' along an invisible axis joining the two fighters, frees them to concentrate on the game and engage within its rules.

The *Virtua Fighter 2.1* update (coin-op and Saturn) accentuates and refines the restriction of freedom of movement in order to heighten the intensity of combat by making it, 'more difficult to back away from an opponent (the pause between steps is slightly longer)' (Sega Enterprises

1995: 25). Wolf has similarly noted the mismatch of construction and experience and the superfluity of three-dimensionality in terms of game-play. In *Tempest*:

> the Z-axis movement is one-way and does not affect play: the player's point-of-view moves through the tunnel only between levels (thus no steering is even required), and the only other Z-axis movement is that of the objects growing in size as they move up the tunnel to its edge, where the player's character is; the 3-D effect is employed just for show.
>
> (Wolf 1997: 18)

Therefore, provided it is not overt, disintegrated or clumsy, restricting freedom of movement is not necessarily constraining. In certain circumstances, it can enable the game and heighten the experience. The producer of Sega's *Lost World* coin-op notes that the team, 'wanted to make more than a simple exploration game . . . different scenarios will be offered, and to show this movie type of adventure adequately, we needed to lose a certain degree of freedom' (Kumagai 1997: 61).

Early videogame designer Eugene Jarvis discusses how access and exploration are responsible for shaping and enabling the experience:

> You had to figure out what your game was about. What is the essence of the experience I'm gonna give people? It's a hard thing, in the design, to say what can I do? Defender was the first game that scrolled, and then with Robotron, I just stuck the guy on one screen. It was kind of about confine-ment. You are stuck on this screen. There's two hundred robots trying to mutilate you, and there's no place to hide, and you'd better kill them or they're gonna kill you, coming from all sides. It was an incredible sweaty palms experience. It's just the confinement. You are stuck in that room. You can't run down the hallway. You can't go anywhere else. You're just totally focused. A lot of times, the games are about limitations. Not only what you can do but what you can't do. Confining your world and focusing someone in that reality is really important.
>
> (Eugene Jarvis in Herz 1997: 79)

While the concentration of visual representation of space is perhaps inevitable, and Dodge and Kitchin (2001) highlight the problems associated with the visual dominance of much contemporary cyberspace, it is important not to disregard the fact that videogames are multi-sensory experiences. Perhaps the most important recent development in audio has been the incorporation of surround sound capabilities in modern consoles (see Crosby 2002b; Clark 2001a, 2001b, 2001c; Belinkie 1999).

While, since the mid-1990s, consoles have possessed the capability to deliver Dolby Pro Logic audio, the Dolby Digital 5.1 capability of Xbox and PlayStation 2 enables sound designers to utilize positional audio cues to enhance the sense of the spatial coherence and integrity of game-worlds. Positional audio extends the player's spatial awareness and experience beyond the visual. Enemies approaching from behind can be heard even if not seen. The important consequence is that the game space is afforded a greater holism and the player is encouraged to remember that the gameworld persists outside and beyond the window of the screen thereby further reinforcing the reality effect of the game. Tactility, too, is employed by designers to heighten the player's spatial experience. As we shall note in Chapter 8, while the vibrating 'Rumble-Paks' incorporated into standard console joypads have been most commonly used to reinforce collisions, explosions or the discharge of weaponry, designers have also begun to experiment with haptic feedback to alert players to oncoming assailants. The ability to *feel* the distant footsteps of an unseen character again broadens the player's sensory experience of – and, it can be argued, their sense of presence *in* – the gameworld. Videogame spaces are experienced viscerally with the whole body. The exploration of videogame space is a kinaesthetic pleasure. It is important, therefore, to consider the ways in which players virtually exist within these spaces. The following chapter presents a discussion of the complex composition of the videogame character and multifaceted relationships between players and their on-screen avatars.

8

VIDEOGAME PLAYERS AND CHARACTERS
Narrative functions and feeling cyborgs

THE VIDEOGAME CHARACTER AS CULTURAL ICON

Sonic the Hedgehog, Super Mario, Lara Croft, Pikachu, Snake Pliskin, Crash Bandicoot, Donkey Kong . . . over 40 years, videogames have given birth to many memorable characters and, while not everybody could identify *The Legend of Zelda*'s Ganondorf in an identity parade or name all 151 original *Pokémon*, Super Mario's beaming face is recognized the world over (see Choquet 2002). Certainly, the increased popularity of videogame play must be partly responsible for the visibility of these characters, however it is notable that recognition extends beyond players. Reporting the findings of a 1990 study (predating the boom in popularity that accompanied the 1994 launch of PlayStation), Sheff (1993: 9) notes that 'the Nintendo mascot, Super Mario, was more recognized by American children than Mickey Mouse'. It is clear that videogame characters have long since broken free of the PC or console screen and, from the days of *Pac-Man* and even *Space Invaders*, their presence has been by no means restricted to the interactive screen (see Kent 2001 on 'Pac-Mania' and 'Space Invaders fever', for example). As such, even those who have never played *Tomb Raider* are likely to have at least heard of Lara Croft.

However, this should probably surprise us little and videogame characters must be seen as treading a similar path to other fictional characters that have 'transcended' the texts of their original appearance (see Denning 1987, for example). In their discussion of 'the Bond

phenomenon', Bennett and Woollacott (1987: 6) point to the impossibility and futility of conceiving the central character as solely constituted
within Ian Fleming's novels or even their film adaptations and posit
James Bond as a popular hero 'constituted within a constantly moving
set of inter-textual relations'. Certainly, the intertextuality of videogame
characters is evident. While videogames may have given them life,
the Pokémon, like Earthworm Jim, the Super Mario Bros and Sonic
the Hedgehog, star in their own television and film series (see Kinder
1991), while Lara Croft publicizes energy drinks and stars in an
action movie. Indeed, cinemas have recently played host to a spate of
videogame-related movies including 'Resident Evil' and 'Final Fantasy',
for example, and videogame characters and series are frequently translated to feature-length Japanese animé ('Tekken' and 'Street Fighter II',
see *Manga Official Website* www.manga.co.uk). Even this only scratches
the surface of the intertextual web within which videogames and their
progeny exist and, while space cannot permit an exhaustive list, some
sense of the scope can be gained by considering the comic serializations,
records and even Pikmin plants with which to transform the garden into
videogame space (in this colonization, we note the interesting reversal
of Jenkin's positioning of videogames as surrogates of the disappearing
'real world' spaces of play!). Moreover, the traffic is not merely uni-
directional. Videogames do not simply sit at the perimeter but are
located firmly within the intertextual network of contemporary media.
As such, we note the continued presence of characters such as Mickey
Mouse in videogames, while Bennett and Woollacott (1987) would
doubtless be unsurprised to find a stream of James Bond games (most
notably *Goldeneye*).

It is important to consider that intertextuality ensures that 'the videogame', just like 'the film' is a slippery term in itself and that such
apparently singular, delineated media forms not only exist within a
multi-media context but also that they are supported by a range of other
media texts and forms. Bennett and Woollacott (1987: 9) point to
publicity posters, for example, though we might also consider box art,
instruction manuals, retail displays and perhaps even in-game movie
sequences, in addition to multimedia marketing materials in the study of
videogames. Crucially, as we have seen, for Bennett and Woollacott,
these media working 'alongside' the films and novels (in the case of
Bond) actively cue the audience and encourage consumption in specific
ways. And perhaps it is the fact that some characters are 'flat' – a matter
that we will return to, shortly – which enables them to cross media and
exceed the boundaries of single texts.

THE LIVES OF MARIO

In *Playing with Power*, Marsha Kinder (1991) points to the ways in which Nintendo's *Super Mario Bros 2* attempts to widen its potential audience by incorporating a range of new player-controllable characters. Unlike the original *Super Mario Bros* title, where only Mario was available to players, in the sequel, four characters can be selected. For Kinder (1991), these characters are implemented to appeal to specific target audiences:

> Clearly designed for the expanded audience, 'Super Mario Brothers 2' gives its players four options for identification: for the core audience of males between seven and fifteen, there are Mario and Luigi, veterans of the original 'Super Mario Brothers,' who have the greatest jumping power; for preschoolers, there's Toad, the tiniest figure, who has the least jumping power but the greatest carrying power; and for females, there's Princess Toadstool, who, despite her inferior jumping and carrying power, has the unique ability of floating for 1.5 seconds – a functional difference that frequently leads my son and his buddies to choose her over the others, even at the risk of transgender identification.
>
> (Kinder 1991: 107)

Kinder presents an interesting position here. Reading the characters as sets of appearances, they appear to target certain audiences of players. However, to those players, the characters are not distinguished or identified with in terms of appearance but rather are differentiated in terms of gameplay-affecting characteristics. Princess Toadstool may well be a female character and, as such, may have gender implications for players; but it should be remembered that she exists in a gameworld. What is critically important in this instance is not her represented gender traits but, rather, the unique ability to float for 1.5 seconds. Faced with a gameworld that is comprised of ravines impassable by the other characters with their limited jumping capacity, it is this functionality that usually influences players' character selections. In this way, L.C. Knights' famous critique of character analysis (1933), in which he asked how many children Lady Macbeth had, resonates: is the player truly interested in such depth of psychological detail when their attentions seem so occupied by other, more pressing functional concerns? For Jenkins (1993: 68), Kinder's account problematizes the very notion of 'character' applied to videogames: 'Does this not suggest that traditional accounts of character identification may be inadequate descriptions for the children's relationships to these figures?'. What Kinder presents may be seen as a non-player's 'reading' of the characters. As she notes, her son and friends employ different criteria in their selections and thereby relate to and understand these characters in a quite specific, game-related manner. These players apparently prioritize the capacity and capability of the character rather than seeking identification through appearance and empathy.

In Jenkins' (1993) critique there is no single Super Mario or Princess Toadstool. Videogame characters, and the worlds they inhabit, can be seen to have at least two distinct spheres of existence depending on their presence in interactive/non-interactive media and, it follows, depending on the type of engagement and relationship between them and the audience/player. In non-interactive portions of games such as introductory sequences, cut-scenes or inter-level breaks, and in cross-media representation such as animated cartoon series, the cast of the *Super Mario* series exist as independent 'characters' and can run around, engage in action, and even speak with autonomy from the audience. However, during interactive sequences of the videogame, the individuality and autonomy of character is subsumed to game-specific techniques and capabilities that the player uses, or embodies within the gameworld. Jenkins (1993: 67) notes the changes in 'character' construction as they move between media both gaining and losing traits as the particular form demands. Jenkins notes that cartoon characters moving into videogames are stripped of their traits and reconfigured as gameplay 'cursors':

> [in videogames,] characters play a minimal role, displaying traits that are largely capacities for action ... The character is little more than a cursor which mediates the players' relationship to the story world.
>
> (Fuller and Jenkins 1995: 61)

It is interesting to note that the analysis presented by Jenkins recalls Propp's (1968) discussion of the Russian folktale or Greimas' (1983) subsequent development of the actantial schema. In an influential investigation, Propp indicated that the psychological motivations of characters were insignificant beside their function in driving the narrative. As such, they are important only insofar as they exact some effect and transformation upon the narrative whether through action or the consequences of this action. Within Propp's analytical framework, there are seven distinct characters within the *dramatis personae* or 'spheres of action'. Table 8.1 is adapted from Berger (1997).

These instrumental spheres of action may be clearly identified in certain videogames such as *The Legend of Zelda* and *Super Mario* series, both

Table 8.1 Propp's spheres of action

Villain	whose action disrupts the equilibrium
Donor	provides the hero with objects, information or advice to aid the resolution of the narrative
Helper	assists the hero restoring equilibrium by solving difficult problems
Princess (and her father)	usually the character most threatened by the villain often having to be saved by the hero. The father may set additional tasks and give away the princess to the hero in marriage upon equilibrium being reinstated
Dispatcher	sends the hero on their quest
Hero	involved in the search for a resolution to the disequilibrium and/or fights with the villain
False hero	an eventually unmasked impostor posing as a hero

of which, as Jenkins notes on the previous page, present a narrative space explicitly modelled on the folktale replete with abducted princesses, mysterious villains, and helpers that may take the form of characters such as 'Navi' (the spirit in *Zelda* who aids the player with useful advice on the nature of puzzles or appropriate techniques), or may be abstract, as in the case of *Mario* where information blocks provide a similar function. Moreover, videogames abound with examples of Helpers and Donors. Again, these spheres of action may take an abstract form as is the case with the 'power-up', for example, that adds capabilities and capacities to the player's complement, or may be presented more recognizably as 'characters' as in *Zelda* where many non-playable characters can be seen to exist for no other reason than to pass on particular objects or powers to the player in order that the next task in the search for narrative resolution may be tackled. For Greimas, Propp's classification could be further reduced to three pairs of similarly functional 'actants' thereby stressing their interrelationships.

Mapping these pairs to videogames such as *Super Mario Bros*, *The Legend of Zelda* or *Final Fantasy*, is a comparatively simple task given the centrality of the quest narrative. Table 8.2 shows such a mapping for Nintendo's *Super Mario Sunshine*.

In discussing the issues that accompany the design of the player's in-game character (most typically, the Proppian 'Hero' or Greimas' 'Subject'), Rouse notes their instrumental nature and provides further reinforcement of the notion of character as function in a similar manner to that posited by Propp or Greimas. For Rouse, there are implications for the manner of presentation:

Table 8.2 Actant relationships in *Super Mario Sunshine*

Subject	Object
(Mario/Player)	(Shine Sprites)
Sender	Receiver
(Pianta Judge)	(Isle Delfino Residents)
Helper	Opponent
(Fludd water pack)	(Liquid Mario and minions, and the game space)

There is a popular misconception in game design that gamers want to have main characters with strong personalities for them to control, particularly in adventure and action games. But if one looks at the most popular entries in these genres, one will quickly notice that the player character's personality is often kept to a minimum. Look at Super Mario 64. Though Mario has a fairly distinctive look, what really is his personality? He does not actually have one, leaving him undefined enough for the player to imprint their own personality on him. What about Lara Croft in *Tomb Raider*? Again, a very distinct appearance, a very undefined personality. And if one looks at the space marine in *Doom* or Gordon Freeman in *Half-Life*, one will find no personality whatsoever.

(Rouse 2001: 229–230)

For Rouse, the reason for the lack of 'personality' is clear. A player-character's personality can be a distancing feature if, for example, it continually engages in dialogue that the player finds either out of keeping, or just plain irritating (see also Ryan 2001).

DEVELOPING CHARACTERS

Examining the production processes of videogames is extremely revealing in this respect. The *Pilotwings 64* design process is illustrative:

[character designs] just turned up one day and we immediately started to implement them in the [partially complete] game. There's no story built around the characters as such, but they are very visible in the game and possess different characteristics, the strong burly guy obviously requires a lot more lift but can turn the hang glider faster.

(Gatchell 1996: 64)

Not only do we note a reaffirmation of the centrality of gameplay-affecting qualities in the differentiation of Hero characters in the videogame, and thus their status as Proppian 'functions', but also that the actual designs, the appearances of these characters, need not be well integrated into the design process. In *Pilotwings 64*, the characters that we finally see in the game are literally 'wrapped around' in-game models defined in functional terms. In addition to Proppian functionality, this perhaps recalls once more the notion of 'round' and 'flat' characters. Where 'round' characters, in the famous definition offered by Forster (1927), might be able to confound and surprise the audience as a product of their complexity, 'flat' characters may be summed up in a single sentence and are typically constructed around a single idea or quality. It should be noted that, while the deployment of these terms in popular parlance valorizes round characters, Forster does not posit flat as a derogatory term. And in *Pilotwings 64*, the characters are resolutely flat in a fashion that precisely complements player interaction.

Taking a somewhat different approach than *Tomb Raider*, advertising campaigns for Ubisoft's *Rayman 2* have attempted to highlight the significance of capability and experiential opportunity in the appeal of the videogame and focus attention away from the appearance of the character and onto action that the player can engage in through the character. 'Rayman' is defined not by his appearance or any traits of individuality or autonomy but by his ability (to allow the player) to run, jump, swim . . . In essence, by the flatness of the character:

> No arms, no legs . . . True, but Rayman can do anything (or almost!): jump, swim, loop de loop, climb, scale walls, slide and fly using his hair as a helicopter.
> Rayman will evolve throughout the game and will be given some temporary powers by his friends such as flying helicopter mode, or grabbing onto Purple Lumz, and even progressively increasing the power of his shot!
>
> (*Ubi Soft Entertainment Official Website*)

In discussing the decisions that influenced the design and implementation of the game's characters in the hugely successful *Metal Gear Solid* games, particularly the player-character 'Solid Snake', Hideo Kojima, the series producer, makes an important observation that highlights some of the differences between characters as presented and engaged with during interactive and non-interactive sequences:

> We tried not to give him [Snake] too much character because we want players to be able to take on his role. Snake isn't like a movie star. He's not someone you watch, he's someone you can step into the shoes of. Playing Snake gives gamers the chance to be a hero.
>
> (Kojima 1998: 43)

Videogame play may then be seen as centred on embodied experience with players using the equipment and capabilities that 'Snake', for example, offers them. During interaction, 'Snake', like 'Mario', 'Sonic the Hedgehog' or 'Lara Croft', is a suite of characteristics rather than a character. During the interactive sequences of videogame play, it does not make sense to talk of player-characters as independent entities. There is no 'Mario' or 'Sonic' to the player – there is only 'me' in the game-world and the functionality of the sphere of action via which the game's narrative may be engaged with.

It might shed more light on the nature of the character in gameworlds if we consider the way in which vehicles in a racing game such as *wipEout Fusion*, which obviously are not meant to be human, are differentiated and presented to the player. In common with games such as *Gran Turismo* and *Ridge Racer*, there are no drivers to select in *wipEout*. Rather, the player selects one of a range of anti-gravity racing craft. The craft are differentiated in terms of their acceleration, cornering ability, resistance to attacks, top speed and so on. No one craft excels in all areas. The maintenance of the all-important balance of the game ensures that all craft are approximately equal in overall capability. A typical arrangement sees craft with rapid acceleration hindered by low top speeds, and craft with high top speeds experiencing poor acceleration. These functional differentiations are by no means trivial and greatly affect the way the game plays. For example, narrow, winding tracks with hairpin bends will be rendered significantly more difficult if the player selects a craft with poor cornering. Similarly, a craft with good acceleration rather than high top speed will benefit the player when racing on tracks comprising a series of interconnecting slow corners. Here, the ability to get the craft up to speed quickly out of slow sections rather than relying on a raw top speed that may never be attained is advantageous. The level of sophistication in matching the craft to the particular demands of specific tracks, is clearly attained only through iteration. Repeat play teaches the player the rigours of the courses and encourages them to be reflexive and critical in the consideration of their own style and capability. The 'use' by players of characters, then, operates along the same lines.

PLAYER PREFERENCES

In this way, it is possible to begin to identify longitudinal changes in character preference. It is common that a novice player will, or will be encouraged to, within the game, select a good 'all-round' vehicle. Rather than sacrificing one capacity in one sphere for excellence in another, the 'novice' vehicle will offer average performance in all areas. Such a compromise represents the perfect choice for a player learning the skills, techniques and mechanics of the game. However, as the player becomes more experienced and begins to better understand the demands of the game, the requirements of particular tracks, and wishes to win more races and progress further through the game, their preferences will shift. No longer will the all-rounder be suitable and particular circuits, tournaments or even game types will influence selection. The differentiation of vehicles along these functional lines is readily understood, yet it is possible to see that characters such as Mario or Sonic are differentiated in much the same way. Importantly, this means that the player's character selection and preferences are motivated by the same desires and influenced by the same gameplay-affecting functionality. Reconsidering *Super Mario Brothers 2*, we note that characters are selected according to the ways in which their particular functional, gameplay-affecting traits impact upon, help or hinder the player to complete the task of the level.

One-on-one combat games or 'beat-'em-ups' are perhaps the most immediately obvious examples of games offering considerable player-character choices with many titles offering upwards of 20 or 30 characters to choose from. *Street Fighter II* is one of the most popular and influential of such games. Although it does not offer the same enormous range of playable characters as more recent titles like *Tekken Tag Tournament*, it nevertheless differentiates them in the same manner. Table 8.3 illustrates the comparative strengths and weaknesses of each of the combatants in the game and may be read as a differentiation of potential Proppian Heroes.

It is notable that these characters are differentiated in exactly the same way as the non-humanoid craft of *wipEout*, for example. In this way, following Newman (2002a, 2002b), it is possible, if controversial, to suggest that it may be less important that Chun Li is a female character than that she is incredibly fast – she has high move speed, jump speed and maximum jump power. However, this is compensated for by the lack of attack strength and poor defence. Ask a *Street Fighter II* player to describe Chun Li and they will be likely to include reference to functionality and capability. Thus, Chun Li is fast, able to jump off walls, and possesses some easily executed special moves (rapid button presses rather than convoluted combinations). As such, Chun Li makes an excellent character for the *Street Fighter II* non-adept – even at the risk of transgender identification. However, while it may be possible to claim that Chun Li's representation is secondary to the speed of her movement, pace of her attacks and ease of her special move execution, Newman

Table 8.3 Character comparison chart for (Super Nintendo Entertainment System) *Street Fighter II Turbo*

Fighter	Move speed	Jump speed	Jump power	Jump attack	Defence
M. Bison	3	3	5	3	3
Sagat	2	3	1	3	3
Vega	3	3	5	3	3
Balrog	3	3	5	3	3
Guile	3	3	3	3	3
Blanka	2	4	4	4	4
Ken	3	3	3	3	3
Chun Li	4	4	5	2	2
Ryu	3	3	3	3	3
Zangief	2	2	1	5	5
Dhalsim	1	1	5	3	1
E. Honda	2	2	2	5	3

Source: Reproduced from Nintendo publicity materials, undated.

(2001) has observed that, when presented with a gendered choice of characters, female players tended to select female characters. The observation raises important experience-level questions. Specifically, if female players can be seen to make decisions based on representation rather than the functionality discussed on p. 134, does this not contradict the assertion that Hero characters are differentiated and understood in terms of gameplay-affecting narrative potential? Importantly, while it is often possible to observe gender identification in the character selection process, this selection criterion does not remain constant in the long term. Newman has suggested that character selections based on these representational traits operate only in the short term during the period of acclimatization with a game. Thus, new and inexperienced players are likely to make selections based on identification with characters at the level of appearance while more experienced players select on the basis of information derived from continued play of the game, analysis of the demands of particular opponents or scenarios, and critical reflection on their own strengths, weaknesses and prowess. Taking this micro-history into account, it is not unexpected that female players are seen to initially select female characters. As experience and knowledge of the game increases and, it follows, knowledge of the simulation and the functionality required of a selected Hero, so character selection criteria and preferences are modified.

Ultimately, Newman has noted it is as likely for an experienced female player to select a male character as for an experienced male player to select a female character. In fact, this process of exploring the potential of Hero characters may be seen as an important part of the game itself and may be seen to represent an element of the puzzle, thereby complicating the initial state and adding variety to the game. Players are required to interrogate the game, expose and deduce the rules, understand the simulation or 'think like a computer' as Friedman (2002) would have it. Furthermore, character selections may be based on personal preference (attack versus defence), the need to complete sections of the game that are best, or even only, completed with certain characters (see Kinder's 1991 example on p. 129), or even to manage the challenge by using an unfamiliar or 'difficult' character (one that may not suit their style or whose parameters and potentialities they are not intimately familiar with). By establishing what is essentially a problem-based learning environment, videogames encourage creative thinking and demand problem-solving skills (see Livingstone 2002: 229–233, for example).

Not every videogame, of course, has a central player-character. Games such as *Command and Conquer* do not position the player as any single character. Rather, the player manipulates and directs troops and units around the game space. In *Tetris*, there is no player-character, just a series of falling blocks. If videogames are about embodied experience, what is the player's point of contact with the gameworld? Who is the player in a game of *Tetris* or *Command and Conquer*? Ted Friedman (2002, 1995) presents a fascinating argument that turns attention away from characters towards the simulation of the game itself. In examining 'simulation games' such as *Sim City* and *Civilization II*, Friedman notes that the player does not identify with any particular character, or assume any single role within the game. In *Civilization II*, the player could be seen as 'king, general, mayor, city planner, settler, warrior and priest, to name but a few' (2002). However, rather than simply suggest a hybridized simultaneity of these various roles, Friedman suggests that the player of the simulation game does not see themselves as any one particular character on the screen, but rather as the sum of every force and influence that comprises the game. Players see themselves as the whole screen:

> When you play a simulation game like Civilization II, your perspective – the eyes through which you learn to see the game – is not that of any character or set of characters, be they Kings, Presidents, or even God. The style in which you learn to think doesn't correspond to the way any person usually

makes sense of the world. Rather, the pleasures of a simulation game come from inhabiting an unfamiliar, alien mental state: from learning to think like a computer.

(Friedman 2002)

While Friedman limits the application of this model to the simulation game, it is possible that it may be more widely applicable.

Newman (2002a) has suggested a similar relationship at work in a game like *Tetris*. The player does not have a represented character in the gameworld, nor does the player identify with each block as it falls (the 'blocks' in *Tetris* are correctly known as 'tetraminoes' being comprised of four elements). Rather, what the player relates to is the entire contents of the gameworld. The tetraminoe falling, those already fallen that now comprise part of the terrain of the gameworld, the walls that confine the movement of the tetraminoes, the manipulative possibilities of each tetraminoe – the logic and mechanic of the entire gameworld. It is possible that this holistic approach to the player's perspective may be yet further applicable to videogames that present player-characters. Perhaps the concentration on Mario even as a set of experiential potentials masks the complexity of the player's perspective. Perhaps the manner in which the *Super Mario* player learns to think is better conceived as an irreducible complex of locations, scenario and types of action. Certainly, it is difficult to dislocate Mario the 'character' from Mario World, with its interconnecting pipes, or from running, jumping and puzzling, or even from enemies, adversaries and opponents. In this way, perhaps the very notion of player-character relationships, and characters in locations performing actions and encountering other non-player-characters, still betrays an insensitivity to the experience of play.

By extending Friedman's 'thinking like a computer' concept, it is possible to argue that playing *Gran Turismo 2* is not, despite the rather grand moniker of 'The Real Driving Simulator', to engage with a driving simulator at all. Rather it is to engage with a simulation presented as an exercise in driving fast cars. By learning to 'drive' any given vehicle in *Gran Turismo*, the player is engaging with, exploring and perhaps ultimately mastering, the game's simulation model. To learn to 'drive' a *Gran Turismo* car is to understand the vagaries of the relationship between the controls, the track model, the physics engine and so on. Importantly, *these* controls, *this* track model and *this* particular physics engine. To learn to drive in *Gran Turismo* is to understand the AI of the other 'drivers' or elements in the simulation. For example, all but the most casual of players learns that 'bouncing' off the scenery and, especially, careering into other vehicles is an extremely effective means of taking tight corners. Certainly, it is easier than trying to pick a racing line. Thus, it is commonplace to see players accelerating hard into hairpin bends safe in the knowledge that they can bounce off the group of cars in front without losing too much speed, thereby powering out of the corner ahead of the pack. It follows that a tactic emerges in which players deliberately drop back from leading positions before otherwise difficult corners so as to enact the strategy and regain the lead that might otherwise have been lost in a spin. A knowledge of the simulation allows the player to exact a net gain. In their discussion of engagement with *Super Ghosts and Ghouls*, Green *et al*. (1998) cite one player's recollection of the encounter with the final 'Boss' enemy:

> Well, he's captured you a girl. I don't know whether she's a princess. I'm not really sure about that. He's captured you a girl so you have to go through eight, no seven, seven levels of perplexing mazes and things like werewolves and zombies . . .
>
> (Green *et al.* 1998: 28)

For Green *et al.*, the player's ability to make sense of the game is demonstrated in the understanding of the role of the 'girl'. Appreciating the character as an element within the simulation, the 'girl' has been captured for you – to provide a reason for playing. Examples such as this bring into focus the dialogue between the player and the simulation model. Following this model, to play a videogame may be seen as involving the scrutiny of the parameters of the simulation and the exacting of a performance within it that maximizes the benefit to the player.

EXPERIENCING AT FIRST HAND: BEING AND WATCHING THE HERO

The discussion of videogame characters and the manner in which players engage with them reveals a number of paradoxes. Chief among them is that players frequently report first-hand experience of explicitly mediated representational gameworlds. That is, these videogames make use of second-, third- or varying-person viewpoints and mediation effects such as camera lens flare. Sega's (coin-op) *Scud Race* demonstrates the paradox where players sit in mock-up car chassis, using a steering wheel and pedals to input controls to a gameworld displayed via a 2D screen upon which the car they are driving is visible in its entirety. Similarly, while Kojima talks of creating *Metal Gear Solid*'s Snake as a character into whose shoes the player can step, Snake is visible on screen. Where, then, is the player in this web of relationships? Can they be simultaneously 'in' the car and watching it? (cf. Skirrow 1990: 330).

The idea of a site where human and computer functions are combined is not indigenous only to videogames. Springer highlights the discourse in popular culture that suggests:

> the possibility of human fusion with computer technology in positive terms, conceiving of a hybrid computer/human that displays highly evolved intelligence and escapes the imperfections of the human body. And yet, while disparaging the imperfections of the human body, the discourse simultaneously uses language and imagery associated with the body and bodily functions to represent its vision of human/technological perfection.
>
> (1991: 303)

As Springer notes, the cyborg is a prevalent theme in modern popular culture and is particularly evident in the science fiction of William Gibson and the 'cyberpunk' movement (see also Featherstone and Burrows 1995 and Chapter 7). The notion of cyborg gained many academic followers (Balsamo 1995; Clark 1995; Lupton 1995; Springer 1991; Haraway 1988, for example) and, given its claims to explain possible human–technology hybridizations, it is potentially useful in the study of videogames.

Central to cyberpunk/cyborg discourse is the often explicit contempt that is shown for the body. Springer (1991: 303) notes that 'The word "meat" is widely used to refer to the human body in cyberpunk texts' such as Gibson's (1984) influential *Neuromancer*. Franck (1995: 20) similarly notes the frequency of both fiction and non-fiction writers' references to 'leaving the body behind' and expresses her disgust at the phrases 'meat puppets' and 'flesh cage' (see also Featherstone and Burrows 1995; Cadigan 1991). This relegation of the body to an inert, imperfect but, importantly, perfectible (through prostheses) site, mirrors a 'post-Platonic' strain of thought which views the body as redundant or the object of mistrust and even revulsion (Ihde 1990: 119). However, it is necessary to challenge the notion that the experiences of these systems and the interactions with them, exist solely at the level of subjectified cognition with the body rendered disintegrated and insignificant (see Argyle and Shields 1996; Stone 1991, for example). Sobchack's (1995: 213) essay on technological prostheses reminds us that we do not just *have* a body, we *are* a body (see also Balsamo 1995: 233; Lupton 1995; Morse 1994; Stone 1991).

Considering the body an unreliable mediator of experience, cyberpunk discourse sees technology, as a means of sidestepping this 'faulty component'. Franck points to the kinaesthetic nature of interactions with these technologies:

> VR is very physical. I won't just see images on a flat screen; I will have the feeling of occupying those images with my entire body . . . To see I must move my head. To act upon and do things in a virtual world I must bend, reach, walk, grasp, turn around and manipulate objects.
>
> (1995: 20)

Thus, according to Franck, it is, in actual fact, the responses and inputs of the 'meat' that enable the interface and lock the player into the very heart of the system and experience. Von Woodtke concurs, illustrating the appropriation of even the simplest interface technologies into the bodily sphere of the user:

> You can experience different levels of involvement by simply using a pencil in different ways. For example, take a pencil and change hands – use the pencil in your left hand if you are right handed. Note how your persona relates differently to the pencil in your 'wrong' hand – now there seems to be a boundary between you and this tool. Now take the pencil in your 'normal' hand. Feel the qualities of the pencil – its length, its balance, the quality of the point, the softness of the lead – notice how your ego becomes involved. Use the pencil for a familiar task. Write your name. Notice how the pencil becomes a part of your total organism as you become involved in the task at hand. This also applies to a mouse or any other pointing device as you work with different computer applications. (Even games if you wish.)
>
> (von Woodtke 1994: 11)

It is important, therefore, to avoid consideration of player interaction with characters as a cerebral or mere ephemeral experience. Indeed, videogame interactivity is a powerful experience precisely because it is so 'bodily'.

BEHIND THE VISUAL

An examination of the ways in which players engage with the gameworld through the interface allows us to problematize some of the taken-for-granted visualism prevalent in the academic and developer communities. Bates suggests that:

> First-person games put the camera in the character's head. The player sees what his character sees. In third person, the camera is outside the main

character, usually floating just above and behind but sometimes moving to different positions to provide a better view of the action ... First-person games tend to be faster paced and more immersive. There is a greater sense of being 'in the world' as the player sees and hears with his character. Third-person games allow the player to see his character in action. They are less immersive but help the player build a stronger sense of identification with the character he is playing.

(2001: 48)

This idea privileges representation, however it can be argued that it is the interface – the feel of the game (see also Myers 1990) – that affects the bond between player and gameworld. Many games, such as *Metal Gear Solid 2*, *Super Mario 64*, *Gran Turismo 3* and *wipEout Fusion* have manipulable cameras and allow dynamic shifts between first- and third-person viewpoints. But how important is this? We have discussed already the range of haptic devices now incorporated into games by designers to encourage 'immersion' of players. If viewpoint is the primary mechanism that generates immersive connectivity, though, then should not this dynamism undermine the integrity of the experience and the player's sense of perceived immersion or presence in the gameworld? That it does not is indicative, perhaps, that viewpoint is by no means decisive.

First-hand participation (FHP) is not necessarily contingent on first-person viewpoint. FHP videogames engender a degree of interactive connection with the gameworld that goes far beyond the abstracted 'use' of a system or vicarious identification with and manipulation of an iconic character or world. As a consequence of the perfect tuning, appropriateness, and feel of the interactive interface, the player arguably becomes enmeshed within the feedback loop of the gameworld. Not merely controlling Mario but embodying the character as a set of available techniques and abilities to be deployed:

When you play a game 10,000 times, the graphics become invisible. It's all impulses. It's not the part of your brain that processes plot, character, story. If you watch a movie, you become the hero – Gilgamesh, Indiana Jones, James Bond, whomever. The kid says, I want to be that. In a game, Mario isn't a hero. I don't want to be him; he's me. Mario is a cursor.

(Fullop 1993, cited in Frasca 2001b)

It is clear from our analysis here that videogame characters are susceptible to a variety if interpretations and may subject themselves to a range of possible readings. However, even if we follow the suggestion of Jenkins (1993) and others that the videogame character exists in a very particular form in played videogames, and may be understood in terms of Proppian narrative function or in the manner of Greimas' actants, the intertextuality of these characters remains inescapable. It is important to note that the 'character' of the *Metal Gear Solid* Hero 'Solid Snake' is revealed differently through cut-scenes and inter-level breaks where he may act independently of the player in order to exact particular impact upon the narrative. However, when interactive control is assumed, the player effectively steps into the shoes of Snake. The player almost *is* Snake. But 'Snake' in this sense is not a 'character' with agency and autonomy and is better conceived as a set of capabilities, techniques and capacities that the player can utilize. 'Snake' here is equipment for play; a vehicle through which the player gains access to the gameworld.

Kojima tackled the problem somewhat differently in *Metal Gear Solid 2*, for PlayStation 2. It is only after some time spent playing that the player is made aware that they have not been performing as Snake, the game's supposed lead character. Rather, they have been acting as Raiden. Through subsequent non-playable cut-scenes, the character of Snake can be explored and developed. In this sense, we might observe that the character of Snake oscillates between 'flat' and 'round' in playable and non-playable sequences:

> In MGS2, I wanted to give depth to the Snake character which is very hard to do when the player is playing Snake himself. So I got the idea of distancing the viewer from Snake, to provide a more objective view of him. Thanks to this system, Snake grows in stature. When you play Raiden, who is only a beginner, and then encounter Snake, he suddenly seems more impressive.
>
> (Kojima 2002)

Nevertheless, while some commentators remark on the primacy of engagement with the character at the level of simulation, it is unlikely that players fail to bring *any* investment and understanding of the characters to the game. As we have suggested, this may occur as a consequence of the cueing that Bennett and Woollacott note, one source of which might be box or manual art (see Provenzo 1991), or even the literal designation of certain characters as 'Princess' or 'Villain'. This can reactivate experience and knowledge gained from other texts, or game sequences. Different modes of engagement may affect the way in which characters are conceived by players and constructed by designers, and certainly the interplay between these various 'existences' of videogame characters clearly warrants further research, and needs to be explored alongside the operation of videogames without identifiable player-characters, or even characters at all (Friedman 2002, 1995). Given their diversity of type, a unified theory of videogame characters and player-character-gameworld connection and hybridization may prove unlikely and an important area of future research must investigate the possibly variegated pleasures derived from play by different players. Even if we accept the differentiated model of engagement where characters move between flat and round, the potential pleasures of exploring character development through cut-scenes may provide a mechanism of differentiating motivations for play.

9

SOCIAL GAMING AND THE CULTURE OF VIDEOGAMES

Competition and collaboration on and off screen

THE MYTH OF THE SOLITARY GAMER

It is commonplace, in both popular and academic discourse, to consider videogaming as a solitary activity. We have noted in Chapter 4, for example, that the overwhelming majority of videogame 'effects' studies focus on the lone player and take little or no account of the presence, let alone the influence, of either simultaneous collaborative play or the social contexts that surround and support videogames. Though the consequences of this misunderstanding of videogame play and the culture in which it is located are far-reaching, it is perhaps understandable given the text-centred nature of the methodologies deployed in these studies.

Games such as *Super Mario Sunshine*, *Tomb Raider* or *Metal Gear Solid*, to take a limited range of examples, are all apparently designed for the single player, exploring and battling alone against the enemies, obstacles and spaces of the gameworld. Certainly, the interactive potential of these games appears to be limited by the single joystick command, literally prohibiting the input of more than one player. Moreover, and demonstrated by each of the games above, what Aarseth (1998) has termed the 'man against the environment' theme in which a single character is charged with the task of saving the day and restoring the previously disrupted equilibrium is such a prominent theme in videogames and, indeed, pervades much Western narrative (see Todorov 1977 and Propp 1968, for example), that the game seems to offer no scope for collectivity and collaboration.

The apparently solitary nature of play has been seized upon by detractors of videogames. While certain lone, private activity such as journal writing may be valorized (see Goody and Watt 1968), the videogame has been positioned as an antisocial force, encouraging players to withdraw from society. As Jessen (1995) has noted:

> Serious criticism is levelled at the influence of the medium on children's social relations. It is a common assumption that computer games lead to children becoming socially isolated, all in their separate rooms where they engage in a lone struggle in the artificial universes of the games. In other words, the computer destroys social relations and playing.
>
> (1995: 6)

Such is the concern in the UK that public figures such as Prince Charles have joined the debate, calling for lottery funding for projects to tempt children away from problematic popular culture practices such as videogame play and back to respectable activities such as reading or theatre-going. 'One of the greatest battles we face today', he has claimed 'is to persuade our children away from the computer games and towards what can only be described as worthwhile books' ('Prince battles videogames', BBC News Online 2001). Clearly, videogames are considered to be wholly worthless by those who share the Prince's views. Yet, more than this, videogames are seen by their detractors as not merely responsible for solitary experiences but for isolating ones, too. As a result, they not only *appeal* to loners, but actually *create them*, hence giving rise to the popular conception of videogame fans as reclusive outsiders, distant and disengaged from society, both unwilling and incapable of interacting with others:

> What seems to differentiate the gamer is the absence of friends and alternative leisure opportunities; heavy gamers resort to solitary media for distraction and entertainment. Our evidence is rather limited on this point but, clearly, video games are an activity, which, like watching TV and videos, is something kids prefer to do when they have no other more social options. Family and sibling play is infrequent, mostly involves playing with brothers, and is more frequent in the occasional player groups.
>
> (Kline 1997, cited in Kline 1999: 19)

For many commentators, it appears that videogames are imbued with a quite insidious potency. The power of videogames seems such that players are precluded from incorporating them into their lives in the moderation that it is implied could save the vulnerable from inevitable harm. In this willingness to view games as addictive and drug-like, we must note an equal and somewhat patronizing unwillingness to acknowledge any sophistication in players' use of media. Jessen (1998) notes that, since their introduction in the 1980s, home computers have given rise to widespread concerns that young people would be 'seduced' by them. Sherry Turkle (1984) has provided what may be the apotheosis of this stance claiming that the seductive qualities of computers and games can be found in their presentation of ordered, rule-governed and ultimately, controllable spaces that place the user or player in a central, masterful role. Comparing this with the chaos and fuzziness of the 'real world', Turkle concludes that videogames attract the narcissist in adolescents and play upon the deviance of their development (see also Provenzo 1991). In these terms, though, the charge seems no more applicable to games played with computers than any others, or perhaps even other non-game representational forms.

However, the positioning of the videogame player as social recluse raises fundamental questions and it is necessary to interrogate some of the presuppositions upon which the designation is based. Is it really true that videogaming is an alternative to social interaction? Do the ways in which videogames are actually used and played support the common-sense notion of play as solitary? In a study of videogame uses and gratifications, Sherry et al. present somewhat different findings:

Individuals who spend the longest hours playing were more likely to report playing for Diversion (e.g. 'I play video games when I have other things to do' and 'I play video games when I am bored') and Social Interaction (e.g. 'My friends and I use video games as a reason to get together').

(Sherry et al. 2001: 11–12)

Sherry et al. conclude from this that, while at least part of the pleasure of videogame play can be understood in terms of the displacement of other, perhaps more mundane, activity, it is naive to simply consider videogames as providing a diversion from other people. In fact, they suggest that videogaming is an inherently social activity. Directly contradicting the idea of the solitary player isolated from social contact, they suggest that, 'frequent game play appears to be highly social; perhaps the practice of standing around on the street corner has shifted indoors to video game play' (Sherry et al. 2001: 11–12). It is interesting to note that not only do Sherry et al. suggest that videogame play is, in itself, social, but that engagement in social play is not limited to casual gamers, with heavy players, or 'hardcore gamers' seemingly as likely to engage in non-solitary play. Certainly, research conducted by Funk (1992),

Emes (1997) and Kestenbaum and Weinstein (1985) concludes that the hypothesized link between frequent videogame play, social withdrawal and isolation cannot be supported with current findings (see also Ivory 2001 and Dorman 1997). As such, just as with the effects research investigating the consequences of interacting with violent content, social or psychopathological effects studies do not present a consensus. It follows that the popular perception of the videogame player as an isolated, withdrawn loner is based as much on presupposition and anecdote than on the findings of scholarly study.

Furthermore, it is essential that we broaden our analysis beyond merely the moment of play. Videogames and videogame play do not exist in a vacuum. Even if they are played alone, these texts and the experiences of them are located within a set of interpretive practices. Understanding this videogame culture is key as it highlights the myriad ways in which videogames provide a stimulus for social activity and privileges the complex sets of reading and production activity that explicitly decry the designation of videogaming as trivial or asocial. However, the presentation of the videogame fan as 'nerd' or 'geek' in popular media is revealing. In attempting to explain the vehemence of the reactions to popular media fandom and drawing on Bourdieu (1984, for example), Jenkins (1992) has pointed to the ways in which popular, fan cultures disrupt and resist cultural hierarchies not only in the voracity of their interest in trivial, low texts, but also because fans appear to engage in types of activity that run contrary to interpretative practices preferred by bourgeois culture. As such, by engaging with the objects of their attention and affection in a most intimate manner and eschewing the aesthetic distance and reverence for authorial ownership and authority in texts, these fans' practices can be seen to directly conflict with the dominant aesthetic logic. It follows that these fans are frequently attacked as deviant or perverse readers and are, thereby, marginalized as 'others', positioned as figures of fun. In this way, the popular presentation of the 'computer geek' or 'nerd' as figure of fun may be seen to serve the same purposes as the ridiculed 'Trekkie' stereotypes that Jenkins (1992) discusses.

This chapter deals with the two associated issues of sociality and videogames. Following Jenkins (1992) and Brooker (2002) among others, the discussion will explore the cultures of fandom and explore the ways in which videogames provide a focus for critical discussion, talk and textual production, thereby acting as a pivotal point in the social and cultural lives of many players. First, however, we shall briefly turn our attentions to issues of sociality within videogames in order to expose the impoverished nature of accounts that posit the videogame solely as a site of solitary pleasure.

THE VIDEOGAME AS SOCIAL SPACE

While studies such as Sherry *et al.* (2001) problematize the notion of videogames as solitary or isolating experiences, they are less forthcoming as to the exact nature of social interaction in gaming. In fact, it is possible to identify a variety of types of social interaction and locations either created within games during play, or as a consequence of play. In a study championing the cause of ethnomethodological approaches to the study of games, and clearly highlighting the benefits of studying videogames in their 'natural' context rather than abstracted into research laboratories, Saxe (1994) notes a variety of social interactions that take place during, and as a result of, videogame play:

> On many occasions, at a particularly popular arcade game such as Virtua Fighter and Mortal Kombat, participants (players, spectators) from diverse racial and age backgrounds are all gathered together, sometimes in very cramped quarters, around the same video screen. On this level, the screen play provides an anonymous opportunity for shared play space among individuals who might not normally participate in joint activities.
>
> (Saxe 1994)

Players not only reported significant social networks oriented around and emerging from gaming, but also that these networks were supportive and non-confrontational. For example, players indicated the ways in which they learned from others, and helped others to learn, by sharing information on strategy and technique through talk and observing of the play of others.

This chimes with the position of Loftuş and Loftus (1983) who have noted that 'extrinsic' reinforcement, such as praise and admiration from peers, constitutes a motivation for play. The simple fact is that, just as not all videogames contain violent content, not all videogames are solitary, single-player experiences. Games such as *Gran Turismo* offer two-player racing options where players can compete against each other in real time. One-on-one combat games (generically classified in industry parlance as 'beat-'em-ups') such as *Virtua Fighter*, *Mortal Kombat* (see Saxe 1994 quote above), *Tekken* and the *Dead or Alive* series, are designed, first and foremost, as multiplayer experiences and often present comparatively weak single-player options. Similarly, the recent trend in First-Person Shooter (FPS) games has been the privileging of multiplayer modes almost to the exclusion of single-player features (most obviously *Quake III Arena*, but note also the multiplayer options presented in *Timesplitters 2* and *Halo*, for example), and the critical and commercial success of titles such as *Goldeneye* is a clear indication of the value placed on multiplayer gaming by players.

Indeed, from Steve Russell's (1962) *Spacewar* onward, multiplayer experiences have been a staple of the videogames industry. Although simultaneous play on a single screen is perhaps the most visible expression of multiplayer potential, there are many ways in which players may directly compete either face-to-face or by taking advantage of the network capabilities of PCs and modern consoles. While the presence as standard of four controller ports on consoles such as Sega Dreamcast, Microsoft Xbox and Nintendo GameCube is a relatively recent occurrence, simultaneous multiplayer gaming is by no means novel and from their earliest days, videogame systems have offered at least the potential for at least two-player simultaneity depending upon software. Typically, this simultaneous play is offered via a single display (monitor or television set) on which is presented either a single view of the gameworld (as *Mario Bros* or *Spacewar*, for example) or a 'split-screen' in which separate areas of the display are dedicated to each player (as *Super Mario Kart*, *Gran Turismo 3*, for example). The majority of these multiplayer games are oriented around game modes that pit players in competition with other. Thus, the object of *Virtua Fighter* is to knock the other player's character out (or out of the ring) while *Super Mario Kart* rewards the racer who breaks the chequered flag first having employed whatever nefarious means to beat the opposition. *Mario Kart* and *Gran Turismo* also offer simultaneous multiplayer modes in which the two human players compete within a larger pack of racers. As such, competition can be seen to operate at two levels with players taking on each other and the computer-controlled opponents. In this way, tackling the game clearly requires exploration of the parameters of the simulation both in terms of the handling and performance of the player's character and the AI of the opposition, and is an exercise in learning the strategy and tactics, strengths and weaknesses of the other player and their command and mastery of the simulation.

Issues of privacy and awareness arise in single display multiplayer games as the display is communal and there exists no potential for information to be relayed to each player individually (see Shoemaker 2000). Most obviously, this impacts upon the ability to evade other players as the position and location of all players in the gameworld is constantly revealed though information about player states, such as health and weaponry is also available. Moreover, selections pertaining to strategy and tactics made on screen through menu systems may be evident to other players, thereby exposing one's hand. Videogame designers have sought to tackle the issue of privacy and awareness in different ways both in software and hardware. *John Madden Football* utilizes a menu system that disguises players selections (though from a usability engineering stance, the absence of confirmatory feedback may be seen as problematic, see Nielsen 2000, for example). Sega tackled the issue with their Dreamcast VMU (Visual Memory Unit), a small device that could be attached to each player's controller and which provided an additional, private display via which information could be relayed and selections made. Although the system was not widely deployed by developers, Nintendo offer similar functionality through their GameCube–GameBoy Advance interconnectivity where the handheld console may be used as a controller so that, with specific software, its display may be used to impart information, or even extend the boundaries of the gameworld (Nintendo have shown demonstrations of a pinball-style game in which the ball may 'fall' from the television screen onto the GameBoy Advance display).

Overcoming some of the issues of privacy and awareness, link-up games essentially make use of separate, self-contained systems for each player. Though link-up facilities exist for PlayStation, PlayStation 2 and Xbox, the need for not only multiple consoles, but also multiple displays and usually multiple copies of the game, renders the exercise rather an expensive and unwieldy undertaking and therefore limited in application. More widespread is the link-up facility offered by handheld consoles and especially the GameBoy. *Pokémon* has perhaps been the most commercially successful application of this connectivity. The original *Pokémon* game in fact came in two versions: *Pokémon Blue* and *Pokémon Red*. Importantly, neither version contained all 151 Pokémon and so, to 'catch em all' as the marketing of the game implored, the player required access to both versions. To facilitate this, owners of *Blue* could, via a cable, physically link their GameBoy to that of a *Red* player and battle or trade. As such, while it is tempting to simply read *Pokémon* as a series of combative and confrontational battles, the game encourages negotiation, bargaining and a consideration of the position of others as players attempt to complete their collections, and may be seen to be inherently social in nature.

However, while *Pokémon* may demand social interaction, and games such as *Bishi Bashi Special*, *Mario Party*, *Gran Turismo* and *Goldeneye* all offer explicit opportunity for simultaneous multiplayer engagement, titles such as *Metal Gear Solid*, *Tomb Raider* and *Super Mario 64* still appear to remain resolutely single-player experiences. Certainly, they all say 'for one player' on the box thus cueing the readings of researchers considering only 'the text' and, thereby, quite naturally giving rise to the belief in the potential of these games to act as sites for solitary play only. However, a more sensitive understanding of the ways in which these videogames are actually played, the way these texts are used, reveals a rather different picture. In fact, even ostensibly single-player games are frequently played by more than one person. There are a number of ways this might be seen to happen. For example, players may play in turns – perhaps taking responsibility for one level or one life each before swapping over, or perhaps comparing completion times, number of items collected and so on. Indeed, many videogames from the 1980s offer their multiplayer optionality in this way with players taking each sequence in turn one after the other rather than competing simultaneously as with *Virtua Fighter et al*.

In addition to this 'relay' play, single-player games are often tackled in teams. As such, games such as *Tomb Raider* or *Metal Gear Solid 2* might be tackled by two or more players simultaneously. Clearly, the game only offers the possibility of one player actually controlling; however, it is frequent to find others actively participating yet not interactively controlling 'players'. Perhaps the most common of these team roles is the 'map reading/making co-pilot', although 'puzzle-solvers' and 'look-outs' are also typical (see Newman 2002a). In this way, it is possible to note also that the competitive element of games such as *Tomb Raider* may give rise to collaborative team play. The team of players operates in competition with the simulation rather than with each other as in some multiplayer games such as *Virtua Fighter* (though note also the team vs team 'clan' play modes of games such as the *Quake* series, see also Chapter 4). In this way, *Alea* may be seen to encourage teamworking. Green *et al*. (1998: 30) have highlighted the ways in which 'non-active' players adopt reflective stances that further their understanding of the mechanics and parameters of the simulation, which they refer to as a 'meta-knowledge of the programming principle of this particular game'. They cite the transcribed accounts of two children during a (single-player) play session in which the non-active player takes the opportunity to scrutinize the operation of the simulation and its responses to player actions by observing the performance of the active player, and incorporates this knowledge assimilating with their own experience and modifying their subsequent performance accordingly.

VIDEOGAME CULTURE

The discussion so far has focused on the sociality fostered within videogames as experienced during play and while Microsoft's Xbox Live headset may be seen as an acknowledgement of the importance of the player interaction that surrounds the game, even this facility is offered only during the temporally delimited online session. Consequently, the social interactions we have identified thus far have, to a certain extent, been bounded by the tempo and duration of the game. Thus, players interact and engage with one another while playing, sharing the experience of the game in various ways, either through competition or collaborative exploration. However, to concentrate solely on the period of play is to significantly impoverish the study of videogames. In other words, videogames are about more than just the act and moment of play itself. Quite apart from the fact that it would be incredible to imagine that sociality and interaction would cease upon game over (see, especially, Farley 2000), there is a raft of activity that supports, amplifies and discusses videogames, their use, design and creation. Videogames exist within what Jessen (1998, 1996, 1995, for example) refers to as a children's computer culture. 'Contrary to appearances, the computer and the games are absorbed into the existing children's culture. This happens very much on that culture's own terms – and often in ways that are quite contrary to the interests of the toy market' (Jessen 1995: 6). The claim is not a new one and we note in the work of Jenkins (1992), Lewis (1992) and Brooker (2002) similar participatory cultures of media fandom associated with *Star Wars* and *Star Trek* among other popular texts, and the ways in which these texts may be 'poached', dissected and reassembled, often explicitly filtered through the experiences of other texts.

As we have noted in Chapter 6, the discussion of fans and their interpretive practices not only rehabilitates and sustains characters through the creative reinvention of fan art and literature but also provides a mechanism through which feelings of disappointment, agitation and frustration with videogames may be discussed (Jenkins 1992). Just as Brooker (2002) has noted the considerable discussion surrounding George Lucas' 'betrayal' of *Star Wars* fans with *The Phantom Menace*, so, too, were videogame discussion boards and, especially, Nintendo fansites ringing with the complaints and concerns of those antagonized by pre-release, trade show footage of the GameCube *Legend of Zelda* game and, in particular, its deployment of a cel-shaded, animé style aesthetic popularized in 2001 and 2002 in games such as *Jet Set Radio*, *Cel Damage* and *Auto Modelista*. For some, the comic book visual style with its wide-eyed, boldly coloured characters and environments that harked back to the Laserdisc visuals of *Dragon's Lair*, for example, was wholly at odds with the series that, in previous incarnations, had conformed largely to a *Dungeons and Dragons*-style 'fantasy' aesthetic presently in vogue as a result of the film adaptations of the *Lord of the Rings* trilogy and interestingly referred to as 'realistic' by Zelda fans (see below).

In their responses, commentators and fans presented an often uncomfortable mix of anger, denial and reverence often manifested as faith in Shigeru Miyamoto as infallible auteur. The following extracts of postings to the 'Zelda Guide' fansite are illustrative. For some, the apparent under-use of the technology was an issue:

> I am a diehard Zelda fan, since the first one came out. But I have to say I was disappointed when they changed it to cel shading. Game cube is capable of so much more, I think Zelda would have been far more exciting in 3D.
>
> (posted 12/3/02)

For others the issue concerned the repositioning of the game. Here a clearly impassioned poster articulates the discussion around the hardcore/mass-market gamer delineation:

> I don't understand nintendo's decision. I read an interview with miyamotto and he said they had plenty of different ideas for the game style. I find it really hard to believe this was the best decision. I mean think about it . . . Zelda has always been a serious adventure style game. They made it look like some kids are gonna pop up and start singing Barney songs . . . Nintendo will realize this when it comes out and they see the sales results. I think this decision has turned every hardcore lover of the series away.
>
> (posted 12/3/02; errors in original)

But faith in Miyamoto's vision is retained by some, 'Sure, the fact that it's cell-shaded is wacky, and kind of annoying, but it's Zelda. And any Zelda game has to be good' (posted 11/8/02).

Far from being isolated and incapable of defending their tastes and preferences against the attacks of popular criticism, fans have access to means by which they can vocally defend and share their enthusiasm for their passions. Importantly, these fans, as identified in Jenkins' examination of *Star Trek* or Brooker's discussion of *Star Wars*, represent not only themselves, but act on behalf of, and within, a larger social and cultural community of fans.

It is perhaps unsurprising that embattled videogame fan cultures exist given the widespread public deprecation of the videogame play and players. As Jenkins notes:

> To speak as a fan is to accept what has been labelled a subordinate position within the cultural hierarchy, to accept an identity constantly belittled or

criticized by institutionalised authorities. Yet it is to speak from a position of collective identity, to forge an alliance with a community of others in defense of tastes which, as a result, cannot be read as totally aberrant or idiosyncratic.

(Jenkins 1992: 23)

But Jenkins (1992: 28) sounds a cautionary note. The celebration of fan culture activities as strategies of popular resistance demands contextualization. Fans operate from a position of social weaknesses and marginalization and much of their activity can be seen as a struggle with media industries to reinstate favourite series, for example.

There is a resignation in some of the postings that speaks of a frustration not only with the vagaries of the Zelda visuals or the implicit marketing orientation, but also of a dislocation and an ability to affect the formulation of games and characters that are clearly much-loved, and heavily invested-in, both financially and emotionally:

> Of course as everyone knows the worst thing about this game is the graphics ... I don't think all this complaining will get Nintendo to change their minds because they're obviously dead set on this decision for whatever crazy reason.

(posted to *Zelda Guide* 12/3/02)

The resignation of some of the posters to the Zelda site finds its contrary in the action of fans of the *Mother* RPG series.

The story of *Mother* is a complex one (see *Star.Net* at Starmen.net for a fuller account). Briefly, two titles were released in Japan, of which only *Mother 2* was translated and localized for the US market where it was known as *Earthbound* (consequently, the original *Mother* is colloquially referred to as *Earthbound Zero* outside Japan). Although *Mother 3* went into development, its production was cancelled for unspecified reasons. Keen to see the game come to fruition, fans at 'Starmen.net' aim to petition the developer and publisher and have set about collecting signatures online. Previous petitions have sought the translation of extant games from the series to contemporary platforms (the Earthbound64 petition raised some 10,000 signatures). The stated aim of the *Mother 3* petition is to collect in excess of 30,000:

> We here at Starmen.Net intend to make the hopes of the Earthbound/ Mother fans known. Regulars on Starmen.Net will know that we have done petitions in the past, and that we take them very seriously ... we want to let Nintendo know how much support there is behind this gaming series

... we want to let them know that we're here and we're desperate for
another Mother game.

(*Mother 3 Petition*, Starmen.net)

To support and publicize the cause, fans have produced a variety of
banner adverts for placement on external websites that can be linked
to the petition pages. Such activity foregrounds the complex communicative networks central to media fandom. It is clear also that the
Internet, and particularly the web, have considerably extended the
communicative and discursive potentials of fans and the various interconnected websites, discussion groups and other forums have become
the nexus for fan activity.

The use of the web by fans is, itself, complex. Official websites are
scoured for valuable details. Screenshots from forthcoming titles are
pored over as interfaces or the implications of particular representational
forms are interpreted and analysed so as to deduce information on gameplay. Moreover, the web's potential has been harnessed by commercial
and non-profit journalistic operations. With access to trade shows at
which video demonstrations of works-in-progress, or even playable
previews of incomplete games, such organizations clearly possess valuable information for game fans. By facilitating the delivery of rich
multimedia and, in particular, video, the web has enabled fans to glimpse
into these trade shows and demonstrations. Direct feeds from video
presentations are relayed via websites and made available to a wider
audience for scrutiny. This development is paralleled offline and it is
increasingly common to find dedicated videogaming magazines supplied
with cover-mounted discs containing both playable game demonstrations and video showreel footage. While the addition of playable game
demonstrations and previews heightens their value and distinguishes
them from similar web journals which, for the console market at least,
are incapable of supplying such material, there is a palpable sense in
which the materials have been sanctioned and institutionalized. Preview
levels and demonstration codes released by developers and publishers
at commercially strategic moments certainly help generate buzz and
resonance about a particular product and may serve to activate the
invisible, 'viral' networks of talk that contemporary marketers seek
to access (Rosen 2001).

Marketers have not missed the potency of the playable demonstration
in generating excitement and found ways of capitalizing. Konami, for
example, initially released their playable demonstration of the eagerly
awaited *Metal Gear Solid 2* as a bonus disc accompanying their less
well-anticipated *Zone of the Enders*, for example, ensuring that the die-

hard *MGS* fan more than likely owns a copy of *ZoE*. However, for the fan communities, the web offers information unlikely to be made available on magazine cover discs or official demonstrations. Most notable is clandestine trade show footage; shaky handheld camerawork and muffled sound only add to the sense of mystery. Footage from trade or press demonstrations, whether clandestine or otherwise, is not only highly prized but is endlessly processed among fan communities. Again, discussion of Zelda is telling. While static screenshots revealed the controversial cel-shaded visual style, video sequences showcasing the animation and flow encouraged a reconsideration. Accepting the variegated responses to the aesthetic, the pre-release editorial at fansite *Zelda Guide* assures fans that the artistic decision brings important gameplay benefits and that the game delivers smooth gameplay and a fluid player experience.

SHARING STRATEGY

> The boys often came to a halt, for example, when they could not find a secret door or how to get past a vampire. In this case, the social network helped with ideas, and for a period the exchange of tips was a central element of the social relations in the group of boys mentioned. A good tip was of great value, and was not simply passed on; it was often held back as a secret until the right moment.
>
> (Jessen 1998: 42)

Sharing and trading knowledge about games is an important part of the social interactions that take place among videogame fans. The complexity of videogames and the wealth of secret features in most titles means that information as to the whereabouts of a particular key, the solution to a particular dungeon, the technique for defeating a particular Boss character, is immensely valuable. There are a variety of ways in which this information can be shared. In some instances, it is swapped or traded in exchange for other information – 'if you tell me how to find the secret door in level three, I'll tell you how to defeat the end of level guardian'. It is also, however, absorbed by players playing in groups.

Travel to any videogame arcade and you will doubtless come across a number of apparent bystanders, not playing, but watching others play. Surely, it cannot be much fun going to an arcade to watch other people play videogames? However, while they may be admiring the skill or luck of the player, it is likely that they are not merely watching. Rather, they are learning. Following the technique and strategy of an experienced player provides an ideal opportunity to improve one's own skill. This need not consist only of learning solutions to obvious complexities such as finding out where the key to a locked door is, or the way through a maze, or the solution to a puzzle. We could consider a racing game and note the ways in which players learn cornering technique, braking points and gearing from each other. Again, knowledge and strategy are shared online as well as offline and numerous sites exist that are dedicated to the discussion of individual games, series of titles, genres and so on. Services range from downloadable FAQs (Frequently Asked Questions) and 'hint sheets', detailing the solutions to common problems, through to full-blown 'walkthroughs' that attempt to outline every step required to complete the game. Most walkthroughs even include instructions for accessing all the hidden items and side-quests within the game. In addition to these published materials compiled by individuals and companies, versions of which are also available in printed form on newsstands and through videogame retailers, the Internet proves a rich array of discursive fora within which strategy and technique can be debated and discussed. Asynchronous BBSs (Bulletin Board Services) or real-time chat such as IRC (Internet Relay Chat) facilities enable players to ask questions about problematic sections of games, or answer the questions of others.

Games and strategy are by no means the only topics of discussion among gamers. There is considerable interest in videogame hardware also. For Jessen (1998), computer game culture is as much about the computer as it is the game. Players spend a great deal of time and effort comparing the technical performance of various hardware platforms, and manufacturers and marketers certainly do little to dissuade them. One of the principal selling points of Microsoft's Xbox is its raw processing power. As such, the pages of magazines are filled with discussions of megahertz, RAM, hard drive size and quite meaningless theoretical polygon calculating capabilities. Such marketing tactics are by no means unique to videogaming. These 'hygienic' features are also used to sell desktop PCs as they facilitate easy, if not informative, comparison in a crowded marketplace and provide manufacturers and retailers with a means of differentiating products:

> it doesn't matter that speed measured in megahertz is not only meaningless to the consumer, but that it doesn't really measure computer speed either ... It doesn't matter that consumers have no idea what a hard drive is compared to RAM. What matters is the numbers game: Bigger and better ...
>
> (Norman 1998: 81)

However, not all discussion is futile or the result of marketing sleight of hand. The criticism of gaming interfaces is perhaps the best example and, again, Xbox is a case in point as its joypad has attracted a great deal of criticism for its size, inelegance and placement of control arrays. Unlike the discussion of hygienic features where comparison of processing speeds misunderstands fundamental architectural differences between systems, here comparison is useful, perhaps even essential and encourages players to interrogate those features that make the GameCube joypad easier to use or more flexible than Microsoft's offering. Indeed, Microsoft have released a smaller version of the Xbox controller in light of continued player criticism.

FANS AS MEDIA PRODUCERS

As Jenkins (1992) and Brooker (2002) have noted, fans are not merely consumers of media texts, no matter how avid or dedicated (as Ang's (1985) *Dallas* viewers or Radway's (1984) romance readers, for example) and considerable creativity and effort is expended on the creation of, for example, fan fiction, fan art and fan music. In the realm of videogames fan websites such as *NintendoLand* (nintendoland.com) encourage the production of prose and poetry that embellishes and supports the characters and narratives of popular Nintendo game series including *Super Mario Bros*, *The Legend of Zelda* and *Metroid*. Interestingly, two varieties of fan fiction ('fanfic') are available. The first is quite traditional, in form at least, and sees the embellishment of narrative themes introduced throughout the games, but the second is quite different.

A subsite of the *NintendoLand* fansite dedicated to the *Legend of Zelda* series, also hosts a number of 'interactives'. Here, hypertextual narratives much in the style of adventure books are presented (see Juul 1999 and Chapter 6). Some 'interactives' extend extant narratives, such as *The Search for Koholint* which positions itself as the 'unofficial sequel to *Link's Awakening*'; all are richly intertextual, attempting to synthesize elements, actions, characters and locales from the various Zelda videogames. However, for all their novelty, the degree of branching is quite limited, and the 'interactivity' is frequently restricted to pressing the web browser's 'Back' button in order to reselect the *correct* option. For example:

> A small red book is up in the rafters. Maybe, just maybe that is the one you've been searching for. Maybe it's the one that mentions Koholint. But why would anyone place the book all the way up there? Was there a

dangerous secret locked away in its pages? Who knows? Should you climb up and get it, or not risk the danger of falling and getting bruised?

- 'Sigh, just get to it'
- 'Don't do anything'

[Selecting 'Don't do anything', leads to,] 'Unfortunately, that book was your only hope of ever finding Koholint Island. Too bad.'

[Hitting the 'Back' button to return to the previous page, we now know to select the correct response and try to retrieve the book.]

(extract from *The Search for Koholint*)

Fan art takes a variety of forms and often involves relocating characters in new locales and applying different aesthetic treatments. While this is equivalent to the production of *Star Trek* fan art described by Jenkins (1992), it must be noted that, given the 'virtual' nature of videogame characters, the potential for extratextual readings is limited as there is no 'real' actor with a parallel career or presence beyond this role. Similarly, this limits the videogame fan's collection of materials as there is no possibility of tracking a performer's career. However, a number of important parallels do exist between the videogame fan art displayed at *NintendoLand* and the slash fiction described by Jenkins.

First, videogame fan art appears to provide a space in which women players can redefine characters. The re-presentation of the *Super Mario* series' Princess Peach as both musclebound or, most notably, as hybrid Peach-Xena Warrior Princess illustrates that fan art is one further channel through which resistance can flow (see Chapter 4). Second, like the fanfic, fan art demonstrates intertextuality in its creation and encourages intertextual readings. Examples abound of assemblages of characters from various game series, or even hybridized versions of characters, crossing elements of Link with Pikachu, for example. Particularly notable is an image that shows both Link and Mario in what is clearly Mario's 'world'. Link's caption 'Why do I have this strange feeling that I'm going to save the wrong princess' serves to neatly and ironically highlight the similarity of underlying objectives present in these two experientially different series.

Fan production also encompasses music; for example, some sites host MIDI files of painstakingly transcribed and re-performed videogame themes. Moreover, fans often remix their favourite tunes, extending them, modifying and repositioning their style and thereby melding their authorship with the original. In doing so, the fan-musician like the fan artist or writer of fan fiction, further invests themselves in the text. Perhaps most unexpected, however, given the discussion of changes in the videogames industry and the marginalization of the lone-coder since the late 1980s (see Chapter 3) is the *The Legend of Zelda: The Grand Adventures*. While the user-creation of 'skins' or even levels is well-known and built into the fabric of many First-Person Shooters (FPSs), the creation of an entire game built around, and extending, an extant franchise, is uncommon even in fandom:

> Although not an official title of the Zelda series, this fan created RPG has very much the spirit of a Zelda game. Created by The Ancient Zodiak, this game is being made for all the fans of the series to enjoy, and even take part in. A demo of this game was released in March, and along with the demo came a contest. The winner of the contest won both the admiration of having their design as the official title screen, and they got to be a character in the game. The winner of this contest was 'Elvie'. Another contest will open soon allowing 8 more people a chance to be in the game. This game is still in development and is estimated to release late 2002 – early 2003.
>
> (from *The Legend of Zelda: The Grand Adventures*)

More extreme forms of fan behaviour can be seen in the 'cosplay' or 'costume play' most prevalent in Japan in which fans (or *otaku* as they are know in Japan) dress as their favourite characters. This phenomenon is typical of 'participatory culture'. Probably the most famous example of such acticity is the cult surrounding the film *The Rocky Horror Picture Show* (1973) where some fans dress as characters at screenings of the movie and, for example, throw rice at the screen during the wedding scene. More recently, similar practices have grown around the *Star Wars* saga (see Brooker 2002: 11–16) in which fans not only attend conventions such as 'Otakon', 'Animazement', 'Nekocon', in costume but also pose for photographs to be posted and rated on the web (see also Jenkins 1992).

A practice which has more political economic consequences and, in a sense, may be more palpably resistant, occurs when participatory culture meets with more general computer culture. Swapping and copying software is one example of the way in which players can subvert the wishes of the games industry. Piracy is a serious issue in all forms of digital media and the profit-function of the videogames industry is by no means unique in suffering at the hands of both large- and small-scale illegal game distribution networks. Thus, piracy is by no means a trivial matter. ELSPA (European Leisure Software Publishers' Association) estimates that piracy cost the UK videogames industry some £3 billion in 1999 alone (www.elspa.com, see also ELSPA 2001). Roger Bennett, director general of ELSPA, explains how the statistics are derived:

The way we come to a £3bn figure – that's a retail value by the way – is based on the fact that for every legitimate copy that is sold, there are ten PlayStation titles or other format titles, that are being sold, given away, copied or downloaded. We estimate that, conservatively, even if there are ten people buying or receiving free pirated products, we say that three out of ten would otherwise go and buy that game if piracy didn't exist.

Worldwide, piracy is estimated to account for £13.1 billion in lost revenue ('CD underground', *Edge* 2000: 71–72). Despite continuing technical and legal attempts to combat software piracy, the harsh reality is that it remains widespread. For Jessen (1998), despite the illegality of many of the trades, the swapping of games is one important way in which social networks arise. Groups of friends often club together to buy a range of titles that they copy among themselves. Moreover, they can use their games as capital to swap with other groups, thus increasing their collection of games and widening social networks.

Throughout this chapter, we have seen a variety of ways videogames can, themselves, be seen to be social in nature, encouraging interaction, teamwork and adversarial combat. Moreover, and frequently overlooked by researchers, videogames exist within a wider computer culture. Furthermore, the ways in which players use and absorb gaming into their cultural lives and social networks demonstrate that the popular perception of gaming as a solitary activity is difficult to sustain. As such, social spaces exist both within and around videogames. In general, though, these observations seem to illustrate something more important still about videogames. Far from being the instruments by which a single-mindedly profit-oriented industry can target a market of 'cultural dupes', videogames are intertextual sites within a network of social relations of some complexity. And, indeed, it can be seen that the ways in which these social relations are negotiated sometimes involves imperatives which are resistant to that industry's logic.

10

FUTURE GAMING
Online/mobile/retro

FROM *PONG* TO PLAYSTATION

Attempting to predict the future trajectory of videogames and videogaming is a highly problematic, perhaps even foolhardy, undertaking. The speed of technological change and the unpredictability of development are considerable. Sitting in a late 1970s' living room, playing *Space Invaders* on the Atari VCS (Video Computer System), it would surely have been impossible to predict the near-photorealistic 3D graphical environments experienced in Dolby Digital 5.1 surround sound made possible by Xbox and PlayStation 2, for example. The future of videogames is not difficult to guess merely because of the pace of technological change and the creative potential placed under the fingertips of developers and designers. Indeed, and in stark contrast to such a technologically deterministic stance, consumer take-up and resistance are important factors and are significant in shaping the nature of videogaming as a set of cultural practices.

Virtual reality and online gaming are, perhaps, the most obvious examples of false starts resulting from consumer resistance. While UK developer Virtuality enjoyed some limited success with their novel, headset-based coin-op games in the mid-1990s, VR and its associated paraphernalia have made no impact on the home market. Despite Atari's pre-production experiments with headset and joystick additions for their Jaguar console that did not make it beyond the trade show circuit, and Nintendo's rather more public Virtual Boy console that was released to considerable consumer apathy and to which there are now almost no references on the company's official website, the technologies of VR have simply not impacted upon computer entertainment in the way that proponents such as Rheingold (1991) predicted. Online gaming, too, has met with mixed fortunes and while popular among some PC users, attempts to bring similar connectivity to the mass-market via videogame consoles such as Sega's Dreamcast have been largely unsuccessful to date. However, both Microsoft and Sony remain publicly committed to online connectivity and although their long-term strategies remain closely guarded commercial secrets, it is known that both companies are keen to position their devices as more than merely online gaming platforms. Since its release, Sony have frequently drip-fed tantalizing press releases that indicate the potential for PlayStation 2 to offer multimedia terminal functionality allowing, for example, the downloading of a variety of content including video and audio alongside games (see *PlayStation Europe Official Website* for further information). At this juncture, online console gaming remains very much an unknown quantity and much of its success rests as much upon the take-up of domestic broadband network connections as on the quality of games and developer support.

It should be added that videogame producers cannot rely on technology alone to provide future success; technology alone is insufficient to impress gamers. In fact, nowhere is this fact clearer than when considering *Pokémon*. The original *Pokémon Blue* and *Pokémon Red* games were not based around cutting-edge technology. Rather, running on the GameBoy, their gameworlds were rendered in 2D, in four shades of grey, with limited animation, and with music and sound effects reminiscent of that emanating from 1970s' consoles and home computers. Yet *Pokémon*'s sales figures and the esteem in which the games are held by fans (note player reviews at *Game FAQs*, for example) speak of the way in which the experience transcends the apparent limitations of the host platform. Revisiting Chapter 2 of this book where we encountered the difficulty of defining the videogame as an object of study, it is useful to note the way in which *Pokémon*, like *Tetris*, for example, undermines the efficacy of attempts to understand videogames simply as audio-visual spectacles.

The audio-visual simplicity of games such as *Pokémon* and the apparent immateriality of the representational capabilities of even the updated GameBoy Advance console (that offers what is essentially an early to mid-1990s' home console in portable form) points to the possibility of a divergent future for videogames. The marketing campaigns deployed by Microsoft stress the raw processing, graphical and audio power of Xbox; it is therefore understandably commonplace to think in terms of subsequent generations of videogame hardware offering the potential for bigger, better, faster games, with larger levels, more detailed visualizations, positional audio (see Clark 2001a, 2001b, 2001c, for example), more complex AI and greater interactive potential. Yet, this tells only half the story and one possible videogame future is decidedly backward-looking.

HAVE WE PLAYED THE FUTURE? RETROGAMING AND EMULATION

It is undoubtedly the case that PlayStation 3 will offer audio-visual sophistication orders of magnitude more impressive and flexible than its predecessor, just as PS2 did before. However, the vibrant retrogaming and emulation scenes offer interesting alternatives to the unstoppable march of contemporary videogame technology. Though they are connected and often conflated, retrogaming and emulation are distinct. Retrogaming describes the growing interest in 'vintage' or 'classic' videogaming hardware and software. The fascination with 1970s', 1980s' and, even, early 1990s' 'vintage' videogaming is often expressed in terms of its 'purity'. Here, 'classic' refers not only to the age of the systems and software, but to their status and, particularly, to their perceived emphasis on gameplay over the trappings of presentation and (re)packaging (see *Retrogamer: Classic Video Game Page*, for example). This sentiment is echoed in Nintendo's own 2002/2003 GameBoy Advance marketing which, in the UK at least, positions the platform as dedicated to 'pure gaming'. GameBoy Advance has quickly become a site for reworked versions of 1980s' titles, most notably the *Super Mario* series, perhaps updated, adding new features or new levels to the original formula, thereby extending the game.

While retrogamers are often concerned with obtaining original hardware, retrogaming does not require retro systems. Despite the considerable trade in vintage equipment that is facilitated both offline and online by dedicated retailers and especially by auction websites such as *eBay*, the emergence of emulation software obviates the need for such equipment. Harnessing the power of modern personal computers, emulators are software applications that mimic the technical functionalities and capabilities of other platforms. Thus, by utilizing such applications, PC users may enjoy virtual implementations of gaming platforms such as the Atari VCS, Nintendo Entertainment System (NES) and Sega MegaDrive (also known as Sega Genesis), as well as a variety of coin-op systems, among many others (see *emulation.net* and *MAME: the Official Multiple Arcade Machine Emulator Website*). These applications merely emulate the operation of the particular platform, but players require software to run under these virtual machines. However, as software for the majority of vintage consoles and coin-op machines was distributed in cartridge form or exists on circuit boards bristling with chips, it is impractical if not impossible to purchase or build readers for such media. Instead, emulator users require what are known as ROMs (Read Only Memory, referring to the physical memory chips in cartridges and on circuit boards wherein the game data is stored). ROMs are data dumps from the original cartridges or program boards. A ROM, therefore, contains all of the data from the original media but exists solely in digital form. The emulator application accesses this ROM data via a virtual interface as though a physical cartridge or circuit board was present and attached. As such, and unlike the GameBoy Advance releases of *Super Mario World*, for example, players of vintage games under emulation play with the original code rather than updated or modified conversions. Thus, all quirks, idiosyncrasies and even bugs may be, in theory at least, perfectly reproduced. However, perfect emulation is not guaranteed. The quality of emulation is variable because the production of emulation software is extremely complex, particularly as many console platforms make use of proprietary chips and algorithms, the function and operation of which are often not widely documented.

It must be noted that emulation sits in what can be most charitably described as a legal grey area. The emulation of devices is not illegal in itself. However, the unauthorized ownership and use of ROMs is. As such, most emulators are officially developed as technical programming experiments. However, this is not to say that one cannot legally play ROMs under emulation and a number of games have been placed in the public domain, meaning that the copyright holders have waived their rights to the works. However, it is also suggested that it is possible to play under emulation games that you already own. Therefore, owning a VCS and a copy of the *Space Invaders* cartridge may allow one to play the game under emulation. However, Nintendo, among others, is very clear that this remains an infringement:

> whether you have an already authentic game or not, or whether you have possession of a Nintendo ROM for a limited amount of time, i.e. 24 hours, it is illegal to download and play a Nintendo ROM from the Internet . . . The introduction of video game emulators represents the greatest threat to date to the intellectual property rights of video game developers.
>
> (*Nintendo Company FAQ*)

It is likely also that such vintage games may form the initial portfolio for 'next generation' 3G mobile devices, though here the decision is motivated as much by technolological limitations as aesthetics or retro nostalgia. Certainly, mobile gaming is predicted to make a significant impact on the market and entertainment software is posited as the 'killer application' for next generation mobiles (though the high profile failure of WAP (Wireless Application Protocol) indicates the ability of consumers to confound pundits and the wishes of network operators). Nonetheless, by 2005, Motorola expect wireless services to account for some 32 per cent of the entertainment market, compared with just 8 per cent in 2000 (Björk *et al.* 2002). Moreover, in order to establish and implement platform standards, Motorola, Nokia, Siemens and Ericsson have formed the 'Mobile Games Interoperability Forum'. Yet, interoperability is one area in which the more established videogame hardware manufacturers may steal a march. Most obviously, Nintendo offer connectivity between GameBoy Advance and GameCube while Sony have tentatively discussed the possibility of distributed computing as part of their future PlayStation strategy and are known to be working with IBM and Toshiba in the creation of a processor called 'The Cell' whose various optimized components need not be contained within a single device. Thus, a future PlayStation, for example, not only need never exist as a unit per se but also may be scalable with a specification effectively defined by the user, or perhaps as a product of many players' components. It is unsurprising that there is much research interest in the role of games within ubiquitous computing (UBICOMP) environments.

One further area of interest concerns the development of 'persistent' games. This, perhaps more than any other possible future, may impact upon the form of videogames. Where most videogames cease upon turning off the console and their gameworlds can be considered to be constituted only while the player chooses to engage with them, persistent universes exist in perpetuity. As such, their simulations continue to operate whether or not a player is engaged with them. Typically, persistent universes are presented online and allow many players to simultaneously engage, however, neither of these are pre-requisite, and a number of single-player console games have offered elements of persistence through, for example, real-time context sensitivity. Taking advantage of the host console's built-in clock and calendar, the gameworld in Sega's *NiGHTS*, for example, adapts to reflect the passing of real-world time, presenting particular themes at specific times of the year. More recently, Nintendo's *Animal Crossing* has offered similar real/virtual synchronicity, while the advent of console network connectivity may render possible massively multiplayer online persistent gameworlds.

CONTINUITY

In concentrating only on change, progress and technological advancement, it is tempting to overlook some of the constancies that a consideration of retrogaming reveals. Perhaps the most immediately obvious element that has remained largely unaltered is the videogame controller. While contemporary videogames may seem immeasurably, even unrecognizably, more sophisticated aesthetically, and may offer more complex and varied gameplay, the player, in the home at least, has at their disposal what is essentially the same input device as was attached to consoles more than 20 years old. The thumb pad controller, or 'D-Pad' pioneered in early consoles and handhelds by Nintendo among others (particularly the Game & Watch series), has been refined ergonomically and functionally, but aside from better and more comfortable grips and more arrays of switches, the basic formula has remained essentially unchanged. Perhaps the most important 'revolution' of recent years was the inclusion of 'analogue' controls (in fact, still digital but mimicking analogue properties) offering proportional rather than momentary control. Thus, moving the analogue stick a little makes Mario walk, pushing it further makes him break out into a jog while pushing it further still sees him run. So important was this development, in offering precision control, that videogames such as *Super Mario 64* would not have been possible without its inclusion on the standard control surface. Such was its impact that the analogue stick is now standard fare on modern console and PC joysticks/joypads, not replacing but sitting alongside the 'digital', momentary controls of the 'D-Pad'. However, it is worth noting that analogue control was not new in 1996 when Nintendo made it the centrepiece of the Nintendo 64 joypad and had been commonplace in the 1970s and 1980s. In fact, the Dragon 32 shipped with, and was berated for, analogue joysticks. It seems clear that at least one direction for videogames, both in terms of hardware and software, rests on the scrutiny and reinvention of its own past. However, it will be interesting to consider the ways in which Sony's camera-based, motion recognition 'Eye Toy' interface will be supported by developers and embraced by players.

WHERE NEXT?

Part of the problem with videogame predictions is that they tend to imply mutual exclusivity. Possible futures are frequently presented as binary oppositions:

- Is the future of gaming to be found online, or in next-generation mobile devices?
- Is the future single-player, multiplayer (or even massively multi-player)?
- Will videogames continue to be distributed as complete entities or as episodes, perhaps developed in response to player feedback?
- Will games become longer or shorter? Will 100-hours plus gaming experiences dominate, or will the quick 10-minute blast triumph?
- Will videogames take their influences from film, or will there be a backlash that sees the resurgence of gameplay over graphics?

There is no reason why videogaming cannot develop in all of these areas simultaneously. Currently, for example, retrogaming sits alongside Xbox and PlayStation 2 while the raiding of back catalogues by publishers eager to capitalize on the retro phenomenon gives rise to a GameBoy Advance portfolio that is forward and backward looking in equal measure. Each possible future offers distinctiveness, and the various developments complement each other. Single-player and multiplayer games have co-existed alongside mobile and home console systems for many years. It is foolish and naive to think that videogaming can exist in only one form. In this way, it is possible to suggest, as Friedman (2002) has, that videogames do not constitute a medium at all. For Kay (1984), the computer is a 'metamedium' able to simulate the properties and characteristics of any other, whether real or imaginary. In this way, videogaming can be delivered through a variety of media. As such, the future of videogaming will not be distinguished by its uniformity, but by its diversity.

BIBLIOGRAPHY

Aarseth, E. (1997) *Cybertext: Perspectives on Ergodic Literature* Baltimore and London: Johns Hopkins University Press

Aarseth, E. (1998) *Allegories of Space: the Question of Spatiality in Computer Games* available at http://www.hf.uib.no/hi/espen/papers/space/

Aarseth, E. (2001a) 'What kind of text is a game?' presented at *International GameCultures Conference*, Bristol, 29 June–1 July

Aarseth, E. (2001b) 'Computer game studies, year one' *GameStudies: The International Journal of Computer Game Research* 1 (1) available at http://www.gamestudies.org/

Ackah, W. and Newman, J. (2003) 'Ghanaian Seventh Day Adventists on and offline: problematising the virtual communities discourse' in Karim, K.H. *The Media of Diaspora* London and New York: Routledge, pp. 203–214

Adams, E. (1997) 'Why "on-line community" is an oxymoron' *Gamasutra* (5 December) available at http://www.gamasutra.com/features/designers_notebook/19971205.htm

Adams, E. (1998) 'Games for girls? Eeeeewwww!' *Gamasutra* (13 February) available at http://www.gamasutra.com/features/designers_notebook/19980213.htm

Adams, E. (2001a) 'Replayability, part 1: narrative' *Gamasutra* (21 May) available at http://www.gamasutra.com/features/20010521/adams_01.htm

Adams, E. (2001b) 'Replayability, part 2: game mechanics' *Gamasutra* (3 July) available at http://www.gamasutra.com/features/20010703/adams_01.htm

Adams, P.C. (1992) 'Television as gathering place' *Annals of the Association of American Geographers* 82 (1): 117–135

Allen, R.C. (1985) *Speaking of Soap Operas* Chapel Hill: University of North Carolina Press

Amory A., Naicker K., Vincent J. and Adams, C. (1998) 'Computer games as a learning resource' in Ottmann T. and Tomek, I. (eds) *Proceedings of ED/MEDIA 98*, Vol. 1, Charlottesville, VA: AACE, 50–55

Anders, P. (1998) *Envisioning Cyberspace: Designing 3D Electronic Space* New York: McGraw-Hill

Anderson, C.A. and Dill, K.E. (2000) 'Video games and aggressive thoughts, feelings, and behavior in the laboratory and in life' *Journal of Personality and Social Psychology* 78: 772–790

Anderson, C.A. and Morrow, M. (1995) 'Competitive aggression without interaction: effects of competitive versus cooperative instructions on aggressive behavior in video games' *Personality and Social Psychology Bulletin* 21: 1020–1030

Ang, I. (1985) *Watching 'Dallas': Soap Opera and the Melodramatic Imagination* London and New York: Methuen

Ang, I. (1996) *Living Room Wars: Rethinking Media Audiences for a Postmodern World* London: Routledge

Argyle, K. and Shields, R. (1996) 'Is there a body in the net?' in Shields, R. (ed.) *Cultures of Internet: Virtual Spaces, Real Histories and Living Bodies* London: Sage, pp. 58–69

Asakura, R. (2000) *Revolutionaries at Sony: the Making of the Sony PlayStation and the Visionaries who Conquered the World of Video Games* New York and London: McGraw-Hill

Baker, M. (ed.) (1984) *The Video Nasties: Freedom and Censorship in the Media* London: Pluto Press

Bakhtin, M. (1981) *The Dialogic Imagination: Four Essays* M. Holquist (ed.), C. Emerson and M. Holquist (trans.) Austin: University of Texas Press

Balsamo, A. (1995) 'Forms of technological embodiment: reading the body in contemporary culture' in Featherstone, M. and Burrows, R. (eds) *Cyberspace/Cyberbodies/Cyberpunk* London: Sage Publications

Barlow, J.P. (1990) 'Life in the data cloud: scratching your eyes back in' *Mondo 2000* (2): 44–51

Barthes, R. (1974) *S/Z* R. Howard (trans.) New York: Hill and Wang

Barthes, R. (1992) 'The reality effect' in Furst, L. (ed.) *Realism* Harlow: Longman

Bartle, R. (1996) 'Hearts, clubs, diamonds, spades: players who suit MUDs' *Journal of MUD Research* 1 (1) available at http://www.mud.co.uk/richard/hcds.htm

Bartlett, E. (2000) *So you Want to be a Games Designer?* (originally published in PC Format (UK), November 2000) available at http://www.igda.org/Endeavors/Articles/ebartlett_printable.htm

Bates, B. (2001) *Game Design: the Art and Business of Creating Games* Roseville, CA: Prima Tech (Game Development Series)

Baudrillard, J. (1993) *The Transparency of Evil: Essays on Extreme Phenomena* London: Verso

Belinkie, M. (1999) *Videogame Music: Not Just Kid Stuff* available at the Videogame Music Archive, http://www.vgmusic.com/vgpaper.shtml

Benedikt, M. (1991) 'Cyberspace: some proposals' in Benedikt, M. (ed.) *Cyberspace: First Steps* Cambridge, MA: MIT Press, pp. 119–224

Bennett, T. (1987) 'Texts, readers, reading formations' in Atridge, D., Bennington, G. and Young, R. (eds) *Post-structuralism and the Question of History* Cambridge: Cambridge University Press

Bennett, T. (1990) *Outside Literature* London: Routledge

Bennett, T. and Woollacott, J. (1987) *Bond and Beyond: The Political Career of a Popular Hero* London: Macmillan

Benyon, D. and Murray, D. (1988) 'Experience and adaptive interfaces' *The Computer Journal* 31(5): 465–473

Beram, I. (2001) 'Levels of complexity: a level-design methodology, part one' *Gamasutra* (17 July) available at www.gamasutra.com/resource_guide/20010716/beram_01.htm

Berens, K. and Howard, G. (2001) *The Rough Guide to Videogaming 2002* London and New York: Rough Guides

Berger, A.A. (1997) *Narratives in Popular Culture, Media and Everyday Life* Thousand Oaks, CA: Sage

Besher, A. (1994) *RIM* London: Orbit

Bignell, J. (1997) *Media Semiotics: an Introduction* Manchester: Manchester University Press

Björk, S., Holopainen, J., Ljungstrand, P. and Mandryk, R. (2002) 'Special issue on ubiquitous games: introduction' *Personal and Ubiquitous Computing* 6: 358–361

Bordwell, D. (1989) *Making Meaning: Inference and Rhetoric in the Interpretation of Cinema* Cambridge, MA: Harvard University Press

Bordwell, D. and Thompson, K. (1997) *Film Art: An Introduction* (5th edition) New York: McGraw Hill

Bourdieu, P. (1984) *Distinction: a Social Change Critique of the Judgement of Taste* R. Nice (trans.) Cambridge, MA: Harvard University Press

Bourdieu, P. (1990) *In Other Words: Essays Towards a Reflexive Sociology* Matthew Adamson (trans.) Stanford: Stanford University Press

Bradford, D. (2001) 'Computer game formats 1977–2001' in ELSPA *The Britsoft Book 2002: the Definitive Guide to the UK Leisure Software Industry* The European Leisure Software Publishers Association & Blueprint Marketing Services Ltd

Braun, C. and Giroux, J. (1989) 'Arcade video games: proxemic, cognitive and content analyses' *Journal of Leisure Research* 21: 92–105

Brooker, W. (2002) *Using the Force: Creativity, Community and Star Wars Fans* New York and London: Continuum

Brooks, P. (1982) 'Freud's master plot' in Felman, S. (ed.) *Literature and Psychoanalysis The Question of Reading: Otherwise* Baltimore and London: Yale University Press

Brooks, P. (1984) *Reading for the Plot: Design and Intentions in Narrative* New York: Vintage Books

Brown, J. (1997) 'All-girl Quake clans shake up boys' world' *Wired* (11 November) available at http://www.wired.com/news/news/culture/story/1885.html

Bruckman, A.S. (1992) 'Identity workshop: emergent social and psychological phenomena in text-based virtual reality' MIT Media Lab, Cambridge, MA available at http://www.cc.gatech.edu/~asb/papers/old-papers.html

Buse, P. (1996) 'Nintendo and telos: will you ever reach the end?' *Cultural Critique* 34, 1996: 163–184

C64.com available www.c64.com last accessed October 2003

Cadigan, P. (1991) *Synners* New York: Bantam Books

Caillois, R. (2001) *Man, Play and Games* Meyer Barash (trans.) Urbana, IL: University of Illinois Press

Cairncross, F. (1997) *The Death of Distance: How the Communications Revolution Will Change Our Lives* Boston: Harvard Business School Press

Calica, B. (1998) 'Cutting cut scenes, or how to stay under 5 CDs and still have a fun game' *Gamasutra* (17 July) available at http://www.gamasutra.com/features/game_design/rules/19980717.htm

Calvert, S.L. and Tan, S. (1994) 'Impact of virtual reality on young adults' physiological arousal and aggressive thoughts' *Journal of Applied Developmental Psychology* 15: 125–139

Campbell, J. (1995) 'Adding a spark to videogames' *Electronics World + Wireless World*, June: 458

Carless, S. (1998) 'Punch kick punch: a history of one-on-one beat-'em-ups' *Gamasutra* (24 April) available at http://www.gamasutra.com/features/game_design/19980424/punchkickpunch_01.htm

Case, L (2000) *The Complete Idiot's Guide to Playing Games Online* Indianapolis, IN: Que

Cassell, J. and Henry, J. (eds) (1998) *From Barbie to Mortal Kombat: Gender and Computer Games* Cambridge, MA: MIT Press

Cassell, J. and Jenkins, H. (1998) 'Chess for girls' in Cassell, J. and Jenkins, H. (eds) *From Barbie to Mortal Kombat: Gender and Computer Games*, Cambridge, MA: MIT Press, pp. 2–45

Castells, M. (1996) *The Rise of the Network Society* Oxford: Blackwell

Cavazza, M., Charles, F. and Mead, S.J. (2002) 'Character-based interactive story-telling' *IEEE Intelligent Systems* 17 (4)

'CD underground' (2000) *Edge* (91): 71–72

'Character comparison chart for (SNES) Street Fighter II Turbo' (reproduced from *Street Fighter II: Nintensive Player's Guide*, Nintendo support materials, undated)

Chen, S. and Brown, D. (2001) 'The architecture of level design' *Gamasutra* (16 July) available at www.gamasutra.com/resource_guide/2001 0716/chen_01.htm

'Children now' (2001) *Fair Play* available at http://www.childrennow.org

Choquet, D. (ed.) (2002) *1000 Game Heroes* Köln: Taschen

Chory-Assad, R. and Mastro, D. (2000) 'Violent videogame use and hostility among high school students and college students' paper presented at the 'Violent Video Games and Hostility' panel of the Mass Communication Division of the National Communication Association annual meeting, Seattle, WA, November 2000

Church, D. (2000) 'Abdicating authorship: goals and process of interactive design' *GDC 2000* (Game Developers' Conference), San Jose, Lecture 5403

Clark, A. (2001a) 'Adaptive music' *Gamasutra* (15 May) available at http://www.gamasutra.com/resource_guide/20010515/clark_01.htm

Clark, A. (2001b) 'Audio content for Diablo and Diablo 2: tools, teams and products' *Gamasutra* (15 May) available at http://www.gamasutra.com/resource_guide/20010515/uelman_01.htm

Clark, A. (2001c) 'An interview with Darryl Duncan' *Gamasutra* (15 May) available at http://www.gamasutra.com/resource_guide/20010515/marks_01.htm

Clark, N. (1995) *Rear-view Mirrorshades: the Recursive Generation of the Cyberbody* in Featherstone, M. and Burrows, R. (eds) *Cyberspace/Cyberbodies/Cyberpunk* London: Sage Publications

Classic Gamer Magazine available at http://www.classicgamer.com last accessed October 2003

Cobley, P. (2001a) *Narrative* London: Routledge

Cobley, P. (2001b) 'Analysing narrative genres' *Sign System Studies* 29 (2): 479–502

Cohen, A.K. (1956) *Delinquent Boys: the Culture of the Gang* London: Routledge & Kegan Paul

Cohen, S. (1972) *Folk Devils and Moral Panics: the Creation of the Mods and Rockers* London: MacGibbon & Kee

'Coin operated: die hard arcade' (1996) *Sega Saturn Magazine* (12) October: 86–89

Computer Space (Killer List of Videogames information) available at http://www.klov.com/C/Computer_Space.html last accessed October 2003

Cooper, J. and Mackie, D. (1986) 'Video games and aggression in children' *Journal of Applied Social Psychology* 16: 726–744

'Crackers delight' (2002) *Edge* (114): 6–9

Crang, M., Crang, P. and May, J. (eds) (1999) *Virtual Geographies: Bodies, Space and Relations* London and New York: Routledge

Crawford, C. (1984) *The Art of Computer Game Design* available at http://www.vancouver.wsu.edu/fac/peabody/game-book/Coverpage.html

Crawford, C. (1995) 'Barrels o' fun' *Interactive Entertainment Design* 8 (3) available at http://www.erasmatazz.com/Library.html

Cringley, R.X. (1996) *Accidental Empires: How the Boys of Silicon Valley Make Their Millions, Battle Foreign Competition, and Still Can't Get a Date* London: Penguin Books

Crosby, O. (2002a) *Working So Others Can Play: Jobs in Video Game Development. Part 1: Series Intro and Game Designers* available at http://www.gignews.com/crosby1.htm

Crosby, O. (2002b) *Working So Others Can Play: Jobs in Video Game Development. Part 2: Artists and Sound Designers* available at http://www.gignews.com/crosby2.htm

Crosby, O. (2002c) *Working So Others Can Play: Jobs in Video Game Development. Part 3: Programmers, Other Occupations, and Resources* available at http://www.gignews.com/crosby3.htm

Cubitt, S. (1991) *Timeshift: On Video Culture* London: Routledge

Cumberbatch, G. (1998) 'Media effects: the continuing controversy' in Briggs, A. and Cobley, P. (eds) *The Media: an Introduction* Harlow: Longman

Curtis, P. (1992) 'Mudding: social phenomena in text-based virtual realities' DIAC-92 proceedings available at ftp://ftp.lambda.moo.mud.org/pub/MOO/papers/DIAC92.txt

Danesi, M. (2002) *The Puzzle Instinct: the Meaning of Puzzles in Human Life* Bloomington: Indiana University Press

Darken, R.P. and Sibert, J.L. (1996) 'Navigating large virtual spaces' *International Journal of Human-Computer Interaction* 8 (1): 49–71

Davies, D. (2001) 'Exploring the business side of the business of making games' *Gamasutra* (7 September) available at http://www.gamasutra.com/features/20010907/davies_01.htm

de Certeau, M. (1984) *Heterologies: Discourse on the Other* Brian Massumi (trans.) Minneapolis, MN: University of Minnesota Press

de Certeau, M. (1994) *The Practice of Everyday Life* Berkeley, CA: University of California Press

DEC PDP-1 Handbook available at http://www.dbit.com/~greeng3/pdp1/

Demaria, R. and Wilson, J. (2002) *High Score! The Illustrated History of Electronic Games* Berkeley, CA: McGraw-Hill/Osbourne

Denning, B. (1987) *Mechanic Accents: Dime Novels and Working Class Culture in America* London: Verso

'*Devil May Cry 2* review' (2003) *Edge* (122) April: 92–93

Dibbel, J. (1993) 'A rape in cyberspace' *The Village Voice* (21 December) pp. 36–42. Excerpt also available as 'A rape in cyberspace; or how an evil clown, a Haitian trickster spirit, two wizards, and a cast of dozens turned a database into a society' in Trend, D. (ed.) *Reading Digital Culture* Oxford: Blackwell, pp. 199–213

'Diddy Kong Racing preview' (1997) *Edge* (51) November: 50–53

Digital Press: the Video Game Database available at http://www.digitpress.com/index2.htm last accessed October 2003

Digitiser (1997) 'Response to reader letter', Channel 4 Television (UK), 30 July: 174 (1 of 5)

Dill, K. and Dill, J. (1998) 'Video game violence: a review of the empirical literature' *Aggression and Violent Behavior* 3: 407–428

Dodge, M. and Kitchin, R. (2001) *Mapping Cyberspace* London: Routledge

Dorman (1997) 'Video and computer games: effect on children and implications for health education' *Journal of School Health* 67 (4): 133–138

Dot Eaters, The: Videogame History 101 available at http://www.emuunlim.com/doteaters last accessed October 2003

Downey, J. and McGuigan, J. (eds) (1999) *Technocities* London: Sage

Drotner, Kirsten (2001) *Medier for Fremtiden* Copenhagen: Høst og Søn (extracts available at http://www.game-research.com/statistics)

'East is Eden' (2002) *Edge* (108): 52–63

eBay online auction site (dynamic pricing e-commerce site offering videogame hardware, software, collectibles and other ephemera), available at www.ebay.com last accessed October 2003

Eco, U. (1986) *Faith in Fakes, Essays* London: Secker & Warburg

EDA Archive (Louis Castle session, RealVideo format), available at http://eda.ucla.edu/archive/ram/winter02/castle.ram

Edge Online available at www.edge-online.com

Egli, E.A. and Meyers, L.S. (1984) 'The role of video-game playing in adolescent life: is there a reason to be concerned?' *Bulletin of the Psychonomic Society* 22: 309–312

Eisenberg, R.L. (1998) 'Girl games: adventures in lip gloss' *Gamasutra* (13 February) available at http://www.gamasutra.com/features/19980213/girl_games.htm

ELSPA (2001) *The Britsoft Book 2002: the Definitive Guide to the UK Leisure Software Industry* The European Leisure Software Publishers Association & Blueprint Marketing Services Ltd

ELSPA (2002) *Video Games Spring Fever* (ELSPA media release, 1 March) available at http://www.elspa.com/mediapack/battle.html

Emes, C.E. (1997) 'Is Mr Pac Man eating our children? A review of the effect of video games on children' *Canadian Journal of Psychiatry* 42 (4): 409–414

emulation.net: Dedicated to Emulation on the Macintosh available at www.emulation.net last accessed October 2003

Faber, L. (1998) *Re: Play Ultimate Games Graphics* London: Laurence King Publishing

Farley, R. (2000) 'Game' *M/C: A Journal of Media and Culture* 3(5) available at http://www.api-network.com/mc/0010/game.html

Featherstone, M. and Burrows, R. (1995) *Cyberspace/Cyberbodies/Cyberpunk: Cultures of Technological Embodiment* London: Sage Publications

Fish, S. (1980) *Is There a Text in this Class? The Authority of Interpretive Communities* Cambridge, MA: Harvard University Press

Fling, S., Smith, L., Rodriguez, T., Thornton, D., Atkins, E. and Nixon, K. (1992) 'Video games, aggression, and self-esteem: a survey' *Social Behavior and Personality* 20: 39–46

Forster, E.M. (1927) *Aspects of the Novel (the Clark Lectures)* London: E. Arnold & Co.

Franck, K.A. (1995) 'When I enter virtual reality, what body will I leave behind?' *Architectural Design* 118: 20–23

Frasca, G. (1999) *Ludology Meets Narratology: Similitude and Differences between (Video)games and Narrative* available at http://www.jacaranda.org/frasca/ludology.htm

Frasca, G. (2000) 'Ephemeral games: is it barbaric to design videogames after Auschwitz?' in Eskelinen M. and Koskimaa R. (eds) *CyberText Yearbook 2000* Research Centre for Contemporary Culture, University of Jyväskylä

Frasca, G. (2001a) *Videogames of the Oppressed: Videogames as a Means for Critical Thinking and Debate* Masters Thesis (Georgia Institute of Technology), available at http://www.jacaranda.org/frasca/thesis/

Frasca, G. (2001b) 'Rethinking agency and immersion: videogames as a means of consciousness-raising' essay presented at *SIGGRAPH 2001* available at http://www.siggraph.org/artdesign/gallery/S01/essays.html

Freedman, J.L. (2001) 'Evaluating the research on violent video games' paper presented at *Playing by the Rules: the Cultural Policy Challenges of Video Games Conference* 26–27 October 2001 available at http://culturalpolicy.uchicago.edu/conf2001/papers/freedman.html

Friedman, T. (1995) 'Making sense of software: computer games and interactive textuality' in Jones, S.G. (ed.) *Cybersociety: Computer-Mediated Communication and Community* Thousand Oaks, CA, and London: Sage Publications

Friedman, T. (2002) *Civilization and its Discontents: Simulation, Subjectivity, and Space* available at http://www.game-research.com/art_civilization.asp

Fuller, M. and Jenkins, H. (1995) 'Nintendo® and New World travel writing: a dialogue' in Jones, S.G. (ed.) *Cybersociety: Computer-mediated Communication and Community* Thousand Oaks, CA: Sage Publications

Funk, J.B. (1992) 'Video games: benign or malignant?' *Journal of Developmental and Behavioural Pediatrics* 13: 53–54

Funk, J.B. (1993) 'Reevaluating the impact of video games' *Clinical Pediatrics* 32 (3): 86–90

Funk, J.B. (2001) 'Girls just want to have fun' paper presented at *Playing by the Rules: the Cultural Policy Challenges of Video Games Conference* 26–27 October available at http://culturalpolicy.uchicago.edu/conf 2001/papers/funk2.html

Game FAQs (offering fan authored walkthrough texts, reviews, FAQs (frequently asked questions) and hosting forums for discussing past, present and future videogame releases), available at www.gamefaqs.com last accessed October 2003

'GameCube on the roll' (2002) *Edge* (112) July: 12

Gamespot available at www.gamespot.com last accessed October 2003

Gatchell, D. (1996) 'Nintendo's Ultramen' *Edge* (UK edition) (29) February: 50–64

Gauntlett, D. (1995) *Moving Experiences: Understanding Television's Influences and Effects* Paris and London: John Libbey

Gauntlett, D. (1998) 'Ten things wrong with the "effects model"' in Dickinson, R., Harindranath, R. and Linné, O. (eds) *Approaches to Audiences – A Reader* London: Arnold

Gauntlett, D. (2000) (ed.) *Web.studies: Rewiring Media Studies for the Digital Age* London: Arnold, see also http://www.newmediastudies.com

Genette, G. (1982) *Narrative Discourse* Oxford: Basil Blackwell

Gibson, W. (1984) *Neuromancer* London: HarperCollins

Gibson, W. (1992) *Virtual Light* London: Penguin

Gibson, W. (1996) *Idoru* London: Penguin

Gillespie, A. and Williams, H. (1988) 'Telecommunications and the reconstruction of regional comparative advantage' *Environment and Planning A* 20: 1311–1321

Goody, J. and Watt, I. (1968) 'The consequences of literacy' in Goody, J. (ed.) *Literacy in Traditional Societies* Cambridge: Cambridge University Press

Graetz, J.B. (1981) 'The origin of Spacewar' available at http://www.wheels.org/spacewar/creative/SpacewarOrigin.html (originally published in *Creative Computing*)

'Gran Turismo: the real driving simulator preview' (1998) *Computer & Videogames (C&VG)* (198): 26–27

'Gran Turismo: the real driving simulator review' (1998) *Computer & Videogames (C&VG)* (199): 62–65

Graybill, D., Kirsch, J.R. and Esselman, E.D. (1985) 'Effects of playing violent versus nonviolent video games on the aggressive ideation of aggressive and nonaggressive children' *Child Study Journal* 15 (3): 199–205

Graybill, D., Strawniak, M., Hunter, T. and O'Leary, M. (1987) 'Effects of playing versus observing violent versus nonviolent videogame on children's aggression' *Psychology* 24 (3) 1987: 1–8

Green, B., Reid, J. and Bigum, C. (1998) 'Teching the Nintendo generation? Children, computer culture and popular technologies' in Howard, S. (ed.) *Wired-Up: Young People and the Electronic Media* London: UCL Press

Greimas, A.J. (1983) *Structural Semantics* Lincoln: University of Nebraska Press

Griffiths, M. (1993) 'Are computer games bad for children?' *The Psychologist: Bulletin for the British Psychological Society* 6: 401–407

Griffiths, M. (1997a) 'Video games and aggression' *The Psychologist* 10 (9): 397–401

Griffiths, M. (1997b) Interviewed in 'Videogame violence: the debate that just will not die' *Edge* (46): 68–73

Griffiths, M. (1999) 'Violent video games and aggression; a review of the literature' *Aggression and Violent Behavior* 4 (2): 203–212

Griffiths, M. and Hunt, N. (1993) 'The acquisition, development and maintenance of computer game playing in adolescence: prevalence and demographic indicators' *Journal of Community and Applied Social Psychology* 5: 189–193

Grossman, D. (2001) 'Trained to kill' *Das Journal des Professoren forum* 2 (2): 3–10

'Gun games "not to blame"' (2002) *Digitiser* Newsburst, Channel 4 Television (UK), 12 March: 481 (4 of 7)

'Half a step beyond . . .' (1998) *Edge* (54) January: 33

Hall, S. (1980) 'Encoding/decoding' in Hall, S., Hobson, D., Lowe, A. and Willis, P. (eds) *Culture, Media, Language* London: Hutchinson

Hanson, G.M.B. (1999) 'The violent world of video games' *Insight on the News* 15 (24): 14–17

Haraway, D. (1989) 'A manifesto for cyborgs' in Nicholson, L. (ed.) *Feminism/Postmodernism* New York & London: Routledge

Heckel, P. (1982) *The Elements of Friendly Software Design* New York: Warner Books

Heim, M. (1994) *The Metaphysics of Virtual Reality* New York: Oxford University Press

Herman, L., Horwitz, J., Kent, S. and Miller, S. (n.d.) *History of Videogames* available at www.gamespot.com/gamespot/features/video/hov last accessed October 2003

Hermes, J. (1995) *Reading Women's Magazines* Cambridge: Polity Press

Hermes, J. (2002) 'The active audience' in Briggs, A. and Cobley, P. (eds) *The Media: an Introduction* (2nd edition) Harlow: Longman

Herz, J.C. (1997) *Joystick Nation: How Videogames Gobbled Our Money, Won Our Hearts and Rewired Our Minds* London: Abacus

Hickman, L. (ed.) (1990) *Technology as a Human Affair* New York: McGraw-Hill

Hill, M.H. (1985) 'Bound to the environment: towards a phenomenology of sightlessness' in Seamon, D. and Mugerauer, R. (eds) *Dwelling, Place and Environment: Towards a Phenomenology of Person and World* Dordrecht: Martinus Nijhoff

'Hip or hype?' (1996) *Edge* (28): 56–64

Holquist, M. (1968) 'How to play Utopia: some brief notes on the distinctiveness of Utopian fiction' in Ehrmann, J. (ed.) *Game, Play, Literature* Boston: Beacon, pp. 106–123

Holtzman, S.R. (1994) *Digital Mantras: the Languages of Abstract and Virtual Worlds* Cambridge, MA: MIT Press

Howland, G. (1998a) *Game Design: the Essence of Computer Games* available at http://www.lupinegames.com/articles/essgames.htm

Howland, G. (1998b) *Introduction to Learning in Games* available at http://www.lupinegames.com/articles/introlearn.htm

Huizinga, J. (1950) *Homo Ludens: a Study of the Play Element in Culture* Boston, MA: Beacon Press

Hunter, C.D. (2000) 'Social impacts: Internet filter effectiveness testing – over- and underinclusive blocking decisions of four popular web filters' *Social Science Computer Review* 18 (2): 214–222

Huntington, J. (1989) *Rationalising Genius: Ideological Strategies in the Classic American Science Fiction Short Story* New Brunswick, NJ, and London: Rutgers University Press

IDSA State of the Industry Report 2000–2001 available at http://www.idsa.com/releases/SOTI2001.pdf last accessed October 2003

Ihde, D. (1990) 'Technology and human self-conception' in Hickman, L. (ed.) *Technology as a Human Affair* New York: McGraw-Hill, pp. 125–134

Intellivision Lives! Official Intellivision Classic Videogame Website available at http://www.intellivisionlives.com/ last accessed October 2003

Ivory, J.D. (2001) *Video Games and the Elusive Search for their Effects on Children: an Assessment of Twenty Years of Research* National AEJMC Conference, available at http://www.unc.edu/~jivory/video.html

Jacobi, S. (1996) *The History of Video Games: an Independent Study* available at http://www.digipen.edu/homepages/alumni/1999/Sjacobi/IndStudy.htm

Jenkins, H. (1992) *Textual Poachers: Television Fans and Participatory Cultures* London: Routledge

Jenkins, H. (1993) '"x logic": repositioning Nintendo in children's lives' *Quarterly Review of Film and Video* 14 (4): 55–70

Jenkins, H. (1998) *Complete Freedom of Movement: Video Games as Gendered Play Spaces* available at http://web.mit.edu/21fms/www/faculty/henry3/pub/complete.html. Also published in Cassell, J. and Jenkins, H. (eds) *From Barbie to Mortal Kombat: Gender and Computer Games*, Cambridge, MA: MIT Press, pp. 262–297

Jenkins, H. (2000) 'Art for the digital age' *Technology Review* available at http://www.technologyreview.com/articles/oct00/viewpoint.htm

Jenkins. H. (2001) 'From Barbie to Mortal Kombat: further reflections' paper presented in the Playing by the Rules Conference, Chicago, 26–27 October. Available at http://www.culturalpolicy.uchicago.edu/conf2001/papers/jenkins.html

Jenkins, H. (2002) Interviewed in 'Learning curve: is the academic community finally accepting videogames?' *Edge* April: 54–61

Jensen, P.K. and Scott, H.A. (1980) 'Beyond competition: organizing play environments for co-operative and individualistic outcomes' in Wilkinson, P.F. (ed.) *In Celebration of Play* London: Croom Helm, pp. 296–308

Jessen, C. (1995) *Children's Computer Culture* available at http://www.hum.sdu.dk/center/kultur/buE/articles.html

Jessen, C. (1996) *Girls, Boys and the Computer in the Kindergarten. When the Computer is Turned into a Toy* available at http://www.hum.sdu.dk/center/kultur/buE/articles.html

Jessen, C. (1998) *Interpretive Communities: the Reception of Computer Games by Children and the Young* available at http://www.hum.sdu.dk/center/kultur/buE/articles.html

Johnson, B. (2001) 'Great expectations: building a player vocabulary' *Gamasutra* (16 July) available at www.gamasutra.com/resource_guide/20010716/johnson_01.htm

Johnson-Eilola, J. (1998) 'Living on the surface: learning in the age of global communication networks' in Snyder, I. (ed.) *Page to Screen: Taking Literacy into the Electronic Era* London: Routledge, pp. 185–210

Jones, M.G. (1997) *Learning to Play; Playing to Learn: Lessons Learned from Computer Games* (online) available at http://intro.base.org/docs/mjgames/

Juul, J. (1998) 'A clash between game and narrative' paper presented at the Digital Arts and Culture conference, Bergen, November

Juul, J. (1999) 'A clash between game and narrative' MA Thesis, available at http://www.jesperjuul.dk/thesis

Juul, J. (2000) 'What computer games can and can't do' paper presented at the Digital Arts and Culture conference, Bergen, August, available at http://www.jesperjuul.dk/text/WCGCACD.html

Juul, J. (2001) 'Games telling stories? A brief note on games and narratives', *Game Studies* (1) available at http://www.cmc.uib.no/gamestudies.org/0101/juul-gts/

Kafai, Y.B. (2001) 'The educational potential of electronic games: from games-to-teach to games-to-learn' paper presented in the Playing by the Rules Conference, Chicago, 26–27 October

Kahneman, D. and Tversky, A. (1982) 'The psychology of preferences' *Scientific American* January: 160–173

Karim, K.H. (ed.) (2003) *The Media of Diaspora* London and New York: Routledge

Kasvi, J.J.J. (2000) 'Not just fun and games: Internet games as a training medium' in Kymäläinen, P. and Seppänen, L. (eds) *Cosiga – Learning With Computerised Simulation Games* HUT: Espoo, pp. 23–34

Katz, E. and Lazarsfeld, P.F. (1955) *Personal Influence: The Part Played by People in Mass Communication* New York: Free Press

Kay, A. (1984) 'Computer software' *Scientific American* 251 (3): 52–59

Keen, B. (1987) '"Play it again, Sony": the double life of home video technology' *Science as Culture* 1: 7–42

Keighley, G. (2001) 'The last few hours' included in *Metal Gear Solid 2: Sons of Liberty* (PlayStation 2, PAL version) bonus DVD. Also available at http://gamespot.com/gamespot/features/video/btg_mgs2/

Kent, S.L. (2001) *The Ultimate History of Video Games: from Pong to Pokémon and Beyond – the Story Behind the Craze that Touched our Lives and Changed the World* Roseville, CA: Prima Publishing

Kestenbaum, G.I. and Weinstein, L. (1985) 'Personality, psychopathology, and developmental issues in male adolescent video game use' *American Academy of Child Psychiatry* 24: 329–337

Kinder, M. (1991) *Playing With Power in Movies, Television and Video Games: From Muppet Babies to Teenage Mutant Ninja Turtles* London: University of California Press

King, N. (2000) 'Hermeneutics, reception aesthetics, and film interpretation' in Hill, J. and Gibson, P.C. *Film Studies: Critical Approaches* Oxford: Oxford University Press

Kitchin, R. and Tate, N.J. (1999) *Conducting Research in Human Geography* Harlow: Longman

Klein, M.H. (1984) 'The bite of Pac-Man' *The Journal of Psychohistory* 11: 395–401

Kline, S. (1997) 'Pleasures of the screen: why young people play video games' proceedings of the International Toy Research Conference, Angouleme, France, November

Kline, S. (1999) 'Moral panics and video games' paper presented at the Research in Childhood, Sociology, Culture and History conference (Child and Youth Culture), University of Southern Denmark, Odense

KLOV.com (Killer List of Videogames) available at www.klov.com last accessed October 2003

Knights, L.C. (1933) *How Many Children Had Lady Macbeth? An Essay in the Theory and Practice of Shakespeare Criticism* Cambridge: Gordon Fraser

Kojima, H. (1998) Interviewed in 'Hideo Kojima profile' *Arcade* 1 (1) December: 42–43

Kojima, H. (2002) Interviewed in 'The making of MGS2', FunTV (documentary feature included with *Metal Gear Solid 2: Sons of Liberty* (PlayStation 2, PAL version) bonus DVD)

Koster, R. (1999) *Games as Art* available at http://www.imaginaryrealities.imaginary.com/volume2/issue6/games_as_art.html

Kreimeier, B. (2000) 'Puzzled at GDC 2000: a peek into game design' available at *Gamasutra* http://www.gamasutra.com/features/20000413/kreimeier_01.htm

Kryzwinska, T. (2002) Interviewed in 'Learning curve: is the academic community finally accepting videogames?' *Edge* (109): 54–61

Kuhlman, J.S. and Beitel, P.A. (1991) 'Videogame experience: a possible explanation for differences anticipation of coincidence' *Perceptual and Motor Skills* 72: 483–488

Kumagai, M. (1997) Interviewed in 'Am3' *Edge* (48): 58–63

Lacey, N. (2000) *Narrative and Genre: Key Concepts in Media Studies* Basingstoke: Palgrave

Lalande, A. (1928) *Vocabulaire Technique et Critique de la Philosophie* Paris: Librairie Félix Alcan

Landow, G. (1991) *Hypertext: The Convergence of Contemporary Theory and Technology* Baltimore, MD, and London: Johns Hopkins University

Lang, R. (1985) 'The dwelling door: towards a phenomenology of transition' in Seamon, D. and Mugerauer, R. (eds) *Dwelling, Place and Environment: Towards a Phenomenology of Person and World* The Hague: Martinus Nijhoff

Laurel, B. (1991) *Computers as Theatre* Menlo Park, CA: Addison Wesley

LeDiberder, A. and LeDiberder, F. (1993) *Qui a Peur des Jeux Vidéo?* Paris: Editions La Découverté

Lefebvre, H. (1991) *The Production of Space* Oxford: Blackwell

Legend of Zelda, The: The Grand Adventures available at www.members. shaw.ca/the_zodiak/tgarpg.html last accessed October 2003

Lewis, L.A. (1992) *The Adoring Audience* London: Routledge

Leyland, B. (1996) 'How can computer games offer deep learning and still be fun? A progress report on a game in development' paper presented at ASCILITE 1996 Adelaide, South Australia, 2–4 December. Available at http://www.ascilite.org.au/conferences/adelaide96/papers/14.html

Liquid Narrative Group available at http://liquidnarrative.csc.ncsu.edu/ last accessed October 2003

Livingstone, S. (2002) *Young People and New Media: Childhood and the Changing Media Environment* London: Sage

Loftus, G.R. and Loftus, E.F. (1983) *Mind at Play: the Psychology of Video Games* New York: Basic Books

Lupton, D. (1995) 'The embodied computer/user' in Featherstone, M. and Burrows, R. *Cyberspace/Cyberbodies/Cyberpunk* London: Sage Publications

Lynch, K. (1960) *The Image of the City* Cambridge, MA: MIT Press

MacEachren, A.M. (1995) *How Maps Work: Representation, Visualization and Design* New York: Guildford

McLuhan, M. (1964) *Understanding Media: the Extensions of Man* London: Routledge

McLuhan, M. and Powers, B. (1992) *The Global Village: Transformations in World Life and Media in the 21st Century* Oxford: Oxford University Press paperback

McRobbie, A. and Thornton, S. (1995) 'Rethinking "moral panic" for multi-mediated social worlds' *British Journal of Sociology* 46 (4): 559–573

'Making of Manic Miner, The' (2001) *Edge* (103): 95

'Making of MGS2, The' (2002) FunTV (documentary feature included with *Metal Gear Solid 2: Sons of Liberty* (PlayStation 2, PAL version) bonus DVD)

MAME: the Official Multiple Arcade Machine Emulator Website available www.mame.net last accessed October 2003

Manga Official Website (UK) available at www.manga.co.uk last accessed October 2003

Manninen, T. (2001) 'Virtual team interactions in networked multimedia games – case: "Counter Strike" – multi-player 3D action game' proceedings of PRESENCE2001 conference, 21–23 May, Philadelphia, PA, Temple University

Maroney, K. (2001) 'My entire waking life' *The Games Journal* available at http://www.thegamesjournal.com/articles/MyEntireWakingLife. shtml

Mediascope (1999) *Video Games and their Effects: Issue Briefs* Studio City, CA: Mediascope Press. Available at http://www.mediascope.org/pubs/ ibriefs/vge.htm

Memarzia, K. (1997) *Towards the Definition and Applications of Digital Architecture* School of Architectural Studies, University of Sheffield. Available at http://www.shef.ac.uk/students/ar/ara92km/thesis

Merleau-Ponty, M. (1962) *Phenomenology of Perception* London: Routledge & Kegan Paul

Microsoft Xbox Official Website available at www.xbox.com last accessed October 2003

Modleski, T. (1984) *Loving with a Vengeance: Mass-produced Fantasies for Women* New York: Methuen

Morley, D. (1980) *The 'Nationwide' Audience: Structure and Decoding* London: BFI

Morley, D. (1986) *Family Television: Cultural Power and Domestic Leisure* London: Comedia/Routledge

Morley, D. (1992) *Television, Audiences and Cultural Studies* London: Routledge

Morley, D. and Silverstone, R. (1990) 'Domestic communication – technologies and meanings' *Media, Culture and Society* 12 (1): 35–56

Morse, M. (1994) 'What do cyborgs eat?: oral logic in an information society' *Discourse* 16 (3): 86–123

Mother 3 Petition available at http://starmen.net/petition last accessed October 2003

Muller, J. (1998) 'A game of their own: new computer software aims to entertain and educate pre-teenage girls' *Boston Globe*, 30 March

Murdock, G., Hartmann, P. and Gray, P. (1992) 'Contextualizing home computing: resources and practices' in Silverstone, R. and Hirsch, E. *Consuming Technologies: Media and Information in Domestic Spaces* London: Routledge

Murray, J.H. (1997) *Hamlet on the Holodeck: the Future of Narrative in Cyberspace* New York: The Free Press

Myers, D. (1990) 'Computer game genres' *Play & Culture* 3: 286–301

Nathan, J. (1999) *Sony: the Private Life* London: HarperCollinsBusiness

Newman, J.A. (2001) 'Reconfiguring the videogame player' paper presented at GameCultures International Computer and Videogame Conference, Bristol, 29 June–1 July

Newman, J.A. (2002a) 'The myth of the ergodic videogame' *Gamestudies* available at http://www.gamestudies.org

Newman, J.A. (2002b) 'In search of the videogame player: the lives of Mario' *New Media & Society* 4 (3): 407–425

Newson, E. (1994) 'Video violence and the protection of children' *Journal of Mental Health* 3: 221–226

Nielsen, J. (2000) *Designing Web Usability: the Practice of Simplicity* Indianapolis, IN: New Riders

'Nintendo boss says people could get bored of computer games' (2002) *Annanova* (June) available at http://www.ananova.com/news/story/sm_602417.html?menu=news.technology.gamingnews

Nintendo Company FAQ available at www.nintendo.com/corp/faqs.html last accessed October 2003

Nintendo Company History available at www.nintendo.com/corp/history.html last accessed October 2003

Nintendo Information for Parents available at www.nintendo.com/parents.jsp last accessed October 2003

'Nintendo's Pokémon headache' (1998) *Edge* (55): 14

NintendoLand available at www.nintendoland.com last accessed October 2003

Norman, D.A. (1998) *The Invisible Computer* Cambridge, MA, and London: MIT Press

Novak, M. (1991) 'Liquid architectures in cyberspace' in Benedikt, M. (ed.) *Cyberspace: First Steps* Cambridge, MA: MIT Press, pp. 225–254

'NuGame culture' (1996) *Edge* (31): 56–66

Official Namco Website available at www.namco.co.jp last accessed October 2003

Orosy-Fildes, C. and Allan, R.W. (1989) 'Videogame play: human reaction time to visual stimuli' *Perceptual and Motor Skills* 69: 243–247

Pagán, T. (2001a) 'Where's the design in level design?, Part One' *Gamasutra* (16 July) available at www.gamasutra.com/resource_guide/20010716/pagan_01.htm

Pagán, T. (2001b) 'Where's the design in level design?, Part Two' *Gamasutra* (16 July) available at www.gamasutra.com/resource_guide/20010716/pagan_04.htm

Pearce, C. (2002) 'The player with many faces: a conversation with Louis Castle by Celia Pearce' *Game Studies: the International Journal of Computer Game Research* 2 (2) available at: http://www.gamestudies.org/0202/pearce/

Pedersen, R.E. (2001) *Pedersen's Principles on Game Design and Production* available at http://www.gamedev.net/reference/articles/article1347.asp

Philips, C.A., Rolls, S., Rouse, A. and Griffiths, M.D. (1995) 'Home video game playing in schoolchildren: a study of incidence and patterns of play' *Journal of Adolescence* 18: 687–691

Piaget, J. (1929) *The Child's Conception of the World* London: Routledge & Kegan Paul

Piaget, J. (1951) *Play, Dreams and Imitation, in Children* London: Routledge & Kegan Paul

Pinter, M. (2001) 'Toward more realistic pathfinding' *Gamasutra* (14 March) available at www.gamasutra.com/features/20010314/pinter_01.htm

'Playing the game' (2001) *Edge* (97) supplement, May

PlayStation Europe Official Website available at www.playstation-europe.com last accessed October 2003

Pong-Story: the Site of the First Video Game available at www.pong-story.com last accessed October 2003

Poster, M. (1997) 'Cyberdemocracy: Internet and the public sphere' in Porter, D. (ed.) *Internet Culture* London: Routledge, pp. 201–218

Pretzsch, B. (2000) *Lara Croft and Feminism*, available at http://www.fraue-nuni.de/students/gendering/lara/home.html

'Prince battles video games' (2001) BBC News Online (now BBCi), Entertainment (New Media) (11 July) available at http://news.bbc.co.uk/hi/entertainment/new_media/1433290.stm

Propp, V. (1968) *Morphology of the Folk Tale* Austin: University of Texas Press (original text 1928; first English translation 1958)

Provenzo, E.F. (1991) *Video Kids: Making Sense of Nintendo* London: Harvard University Press

Radway, J. (1984) *Reading the Romance: Women, Patriarchy and Popular Literature* Chapel Hill University of North Carolina Press

'Raising the stakes in the funding lottery' (2002) *Edge* (113): 6–9

'Record year for computer games' (2002) BBC News Online (now BBCi), Entertainment (New Media) (10 January) available at http://news.bbc.co.uk/hi/english/entertainment/new_media/newsid_1752000/1752522.stm

Reid, E. (1995) 'Virtual worlds: culture and imagination' in Jones, S. (ed.) *Cybersociety: Computer-Mediated Communication And Community* Thousand Oaks, CA: Sage Publications

Relph, E. (1976) *Place and Placelessness* London: Pion

Retro (2002) A special edition of *Edge* magazine last accessed October 2003

Retrogamer: Classic Video Game Page available at http://retrogamer.mersey-world.com/ last accessed October 2003

Retro-Games available at http://www.retro-games.co.uk last accessed October 2003

Rheingold, H. (1991) *Virtual Reality* New York: Touchstone

Rheingold, H. (1993) *The Virtual Community: Homesteading on the Electronic Frontier* Reading, MA: Addison-Wesley

Ricoeur, P. (1981) 'Narrative time' in Mitchell, W.J.T. (ed.) *On Narrative* Chicago and London: University of Chicago Press

Roberts, A. (2000) *Science Fiction* London: Routledge

Robins, K. (1995) 'Cyberspace and the world we live in' in Featherstone, M. and Burrows, R. *Cyberspace/Cyberbodies/Cyberpunk* London: Sage Publications

Rollings, A. and Morris, D. (2000) *Game Architecture and Design* Scottsdale, AZ: Coriolis

Rosen, E. (2001) *The Anatomy of Buzz: Creating Word-of-Mouth Marketing* London: HarperCollinsBusiness (see also http://www.emanuel-rosen.com/)

Rosenberg, M.S. (1992) *Virtual Reality: Reflections of Life, Dreams and Technology. An Ethnography of a Computer Society* available at ftp://sunsite.unc.edu/pub/academic/communications/papers/muds/muds/Ethnography-of-a-Computer-Society

Rosenthal, E.M. (1987) 'VCRs having more impact on network viewing, negotiation' *Television/Radio Age* 25 May

Ross, R. (2001) 'Interactive music . . . er, audio' *Gamasutra* (15 May) available at http://www.gamasutra.com/resource_guide/20010515/clark_01.htm

Rouse, R. (2001) *Game Design Theory and Practice* Plano, TX: Wordware Publishing

Ryan, M.-L. (2001) 'Beyond myth and metaphor – the case of narrative in digital media' *Game Studies* (1) available at http://www.cmc.uib.no/gamestudies.org/0101/ryan

Ryan, T. (1999a) 'Beginning level design, Part 1: level design theory' *Gamasutra* (16 April) available at www.gamasutra.com/features/19990416/level_design_01.htm

Ryan, T. (1999b) 'Beginning level design, Part 2: rules to design by and parting advice' *Gamasutra* (23 April) available at www.gamasutra.com/features/19990423/level_design_01.htm

Sabin, R. (1998) 'Eurocomics: "9th art" or misfit lit?' in Briggs, A. and Cobley, P. (eds) *The Media: an Introduction* Harlow: Longman

Sanger, G.A. (2001) 'The sound of money (down the potty): common audio mistakes in kids' games' *Gamasutra* (15 May) available at http://www.gamasutra.com/resource_guide/20010515/sanger_01.htm

Sardar, Z. (1995) 'alt.civilisations.faq: cyberspace as the darker side of the West' *Futures* 27: 777–794

Saxe, J. (1994) 'Violence in videogames: what are the pleasures?' paper presented at the International Conference on Violence in the Media St John's University, 3–4 October. Available at http://www.media-awareness.ca/eng/issues/violence/resource/reports/gamedoc.htm

SCEI (Sony Computer Entertainment Inc.) (2002) *PlayStation®2 Achieves Record Sales of 5 Million Units Worldwide in the Holiday Season* (SCEI press release) available at http://www.scei.co.jp/corporate/release/index.html

Schramm, W., Lyle, J. and Parker, E.B. (1961) *Television in the Lives of Our Children* Stanford, CA: Stanford University Press

Schroeder, R. (1994) 'Cyberculture, cyborg post-modernism and the sociology of virtual reality technologies: surfing the soul of the information age' *Futures* 26: 519–528

Schuler, D. (1996) *New Community Networks: Wired for Change* New York: ACM/Addison Wesley

Schutte, N.S., Malouff, J.M., Post-Gorden, J.C. and Rodasta, A.L. (1988) 'Effects of playing video games on children's aggressive and other behaviors' *Journal of Applied Social Psychology* 18: 454–460

Scott, D. (1995) 'The effect of video games on feelings of aggression' *The Journal of Psychology* 129 (2): 121–132

Seamon, D. and Mugerauer, R. (eds) (1985) *Dwelling, Place and Environment: Towards a Phenomenology of Person and World* Dordrecht: Martinus Nijhoff

Search for Koholint, The available at www.nintendoland.com/zelda/kokiri/interact/index.htm last accessed October 2003

Seay, J. (1997) *Education and Simulation/Gaming and Computers* available at http://www.cofc.edu/~seay/cb/simgames.html

Sega Enterprises (1995) *Virtua Fighter 2* Instruction manual and packaging (PAL-Europe version)

Segal, K. and Dietz, W. (1991) 'Physiological responses to playing a video game' *American Journal of Diseases of Children* 145 (9): 1034–1036

Seiter, E., Borchers, H., Kreutzner, G. and Warth, E. (eds) (1989) *Remote Control: Television, Audiences and Cultural Power* London: Routledge

Serious Fun (*Equinox* series) Uden Associates for Channel 4 Television (UK). First broadcast 7 p.m. 5 December 1993

Sheff, D. (1993) *Game Over: Nintendo's Battle to Dominate an Industry* London: Hodder and Stoughton

Sheppard, A. (1997) 'Social learning and video games' (letter) *The Psychologist* November: 488–489

Sherman, B. and Judkins, P. (1992) *Glimpses of Heaven, Visions of Hell: Virtual Reality and its Implications* London: Hodder & Stoughton

Sherry, J., Lucas, K., Rechtsteiner, S., Brooks, C. and Wilson, B. (2001) 'Video game uses and gratifications as predictors of use and game preference' paper presented at the ICA convention Video Game Research Agenda Theme Session Panel, 26 May 2001. Available at http://www.icdweb.cc.purdue.edu/~sherry/videogames/VGUG.pdf

Shields, R. (ed.) (1996) *Cultures of Internet: Virtual Spaces, Real Histories, Living Bodies* London: Sage

Shoemaker, Garth B.D. (2000) 'Privacy and awareness in multiplayer electronic games' in proceedings of the Western Computer Graphics Symposium. Panorama Mountain Village, Canada, March 2000, pp. 117–121

Shuker, R. (1995) 'Game far from over: the video game phenomenon' (originally in *SCRIPT*, (New Zealand) *Journal of the National Association of Teacher Educators* 34). Available at http://www.massey.ac.nz/~wwedpsy/tandsres/roy1.htm

Siegel, A.W. and White, S.H. (1975) 'The development of spatial representation of large-scale environments', in Reese, H.W. (ed.) *Advances in Child Development and Behavior* New York: Academic Press

Silverstone, R. (1994) *Television and Everyday Life* London: Routledge

Skelton, T. and Valentine, G. (1998) *Cool Places: Geographies of Youth Cultures* London: Routledge

Skirrow, G. (1990) 'Hellivision: an analysis of videogames' in Alvarado, M. and Thompson, J.O., *The Media Reader* London: BFI Publishing (originally published in MacCabe, C. (1986) *High Theory/Low Culture* Manchester: Manchester University Press)

Smith, H. (2001) *The Future of Game Design: Moving Beyond Deus Ex and Other Dated Paradigms* available at http://www.igda.org/Endeavors/Articles/hsmith_printable.htm

Smith, J.H. (2001a) 'The forgotten medium' *Game Research* available at http://www.game-research.com/art_the_forgotten_medium.asp

Smith, J.H. (2001b) 'What women want – (and it ain't Counter Strike)' *Game Research* (17 November) available at http://www.game-research.com/art_what_women_want.asp

'SNK rolls out 64-bit hardware' (1997) *Edge* (48) August: 12

Sobchack, V. (1995) 'Beating the meat/surviving the text, or how to get out of this century alive' in Featherstone, M. and Burrows, R. (eds) *Cyberspace/Cyberbodies/Cyberpunk* London: Sage Publications

Spacewar! (Java version of Steve Russell's PDP-1 'Spacewar' game) available at http://lcs.www.media.mit.edu/groups/el/projects/spacewar/ last accessed October 2003

Springer, C. (1991) 'The pleasure of the interface' *Screen* 32 (3): 303–323

'Star Wars: Episode II preview' (2002) *Digitiser* (Game 2002 preview), Channel 4 Television (UK) Teletext, 12 February: 483

Star.Net: Get Back in the Game available at http://starmen.net last accessed October 2003

Stephenson, N. (1992) *Snow Crash* New York: Bantam Spectra

Sterling, B. (1996) *Holy Fire* New York: Bantam Books

Stone, A.S. (1991) 'Will the real body please stand-up? Boundary stories about virtual cultures' in Benedikt, M. (ed.) *Cyberspace: First Steps* Cambridge, MA: MIT Press, pp. 81–118

Stout, B. (1997) 'Smart moves: intelligent pathfinding' *Gamasutra* (August) available at www.gamasutra.com/features/19970801/pathfinding.htm

'*Super Mario 64* review' (1996) *Maximum* (7): 116–117

Sutton-Smith, B. (1997) *The Ambiguity of Play* Cambridge, MA, and London: Harvard University Press

Takahashi, D. (2002) *Opening the Xbox: Inside Microsoft's Plan to Unleash an Entertainment Revolution* Roseville, CA: Prima Publishing

Thompson, J.B. (1990) *Ideology and Modern Culture: Critical Social Theory in the Age of Mass Communication* Cambridge: Polity

Tivers, J. (1996) 'Playing with nature' *The Geographical Magazine*, April: 17–19

Todorov, T. (1977) *The Poetics of Prose* Ithaca, NY: Cornell University Press

'Top ten industry facts' (2002) IDSA (Interactive Digital Software Association) available at http://www.idsa.com/pressroom.html, last accessed April 2002

Tuan, Yi-Fu (1974) *Topophilia: a Study of Environmental Perception, Attitudes and Values* Englewood Cliffs, NJ: Prentice-Hall

Tuan, Yi-Fu (1977) *Space and Place: The Perspective of Experience* London: Edward Arnold

Turkle, S. (1984) *The Second Self: Computers and the Human Spirit* New York: Simon & Schuster

Ubi Soft Entertainment Official Website available at www.ubi.com last accessed October 2003

'US sales go big' (2002) *Digitiser*: Newsburst, Channel 4 TV (UK), 12 February: 481

'Vibrating games health warning' (2002) BBC News Online (now BBCi), Health (12 February) available at http://news.bbc.co.uk/1/hi/health/1792102.stm

Vidart, D. (1995) *El juego y la condición humana. Notas para una antropolgía de la libertad en la necesidad* Montevideo: Ediciones de la Banda Oriental

'Video nasties' (*Dispatches* series) (2000) 3BM for Channel 4 Television (UK). First broadcast 9 p.m. 23 March

Videotopia A travelling museum exhibition dedicated to coin-ops. Details available at www.videotopia.com last accessed October 2003

'Vintage gaming' (2002) *Edge* (107) February: 64–69

Vogler, C. (1998) *The Writer's Journey: Mythic Structure for Storytellers and Screenwriters* London: Pan

von Woodtke, M. (1994) 'Hand to mouse: the mental connection' (Keynote Address, 6th Annual Conference of the Design Communication Association) *Representation* (Journal of the Design Communication Association): 9–11

Waggoner, B. and York, H. (2000) 'Video in games: the state of the industry' *Gamasutra* (3 January) available at http://www.gamasutra.com/features/20000103/fmv_01.htm

Wall, J. (2002) 'Music for *Myst III*: Exile – the evolution of a videogame soundtrack' *Gamasutra* (15 May) available at http://www.gamasutra.com/resource_guide/20010515/clark_01.htm

Warne, P. (2001) 'Three inspirations for creative level design' *Gamasutra* (16 July) available at www.gamasutra.com/resource_guide/20010716/warne_01.htm

Waters, D. (2002) '*Metal Gear 2* lacks solidity', BBC News Online (now BBCi) (8 March) available at http://news.bbc.co.uk/hi/english/entertainment/reviews/newsid_1857000/1857454.stm

Waugh, A. (1984) *Metafiction* London: Methuen

Wellman, B. and Gulia, M. (1999) 'Virtual communities as communities: net surfers don't ride alone' in Smith, M.A. and Kollock, P. (eds) *Communities in Cyberspace* London: Routledge, pp. 167–194

Whitlock, T.D. (1994) *Fuck Art, Let's Kill! Towards a Post Modern Community* available at http://www.fed.qut.edu.au/tesol/cmc/emu/moo_papers.html

Williams, R. (1974) *Television: Technology and Cultural Form* London: Fontana

Witmer, B.G., Bailey, J.H., Knerr, B.W. and Parsons, K.C. (1996) 'Virtual spaces and real-world places: transfer of route knowledge' *International Journal of Human-Computer Studies* 45: 413–428

Wolf, M.J.P. (1997) 'Inventing space: toward a taxonomy of on- and off-screen space in video games' *Film Quarterly* 51 (1): 11–23

Woodward, K. (1994) 'From virtual cyborgs to biological timebombs: technocriticism and the material body' in Bender, G. and Druckrey, T. (eds) *Culture on the Brink: Ideologies of Technology* Seattle, WA: Bay Press

Wurtzel, J. (2001) 'Games blur fantasy and reality', BBC News Online (now BBCi), Science & Technology (27 August) available at http://news.bbc.co.uk/hi/english/sci/tech/newsid_1507000/1507296.stm

Xbox Linux Project available at http://xbox-linux.sourceforge.net/ last accessed October 2003

'Yoshi's story review' (1998) *Computer & Video Games* (197) April: 46–49

Young, R.M. (2000) 'Creating interactive narrative structures: the potential for AI approaches' in the working notes of the AAAI Spring Symposium on Artificial Intelligence and Interactive Entertainment, Stanford, CA, March 2000

Young, R.M. and Riedl, M. (2003) 'Towards an architecture for intelligent control of narrative in interactive virtual worlds' presented at *IUI'03*, Miami, FL: 12–15 January

Zelda Guide: Your Guide to Zelda on the Net available at www.zeldaguide.com last accessed October 2003